THE ATLAS OF THE SOLAR SYSTEM

THE ATLAS OF THE
SOLAR SYSTEM

BILL YENNE

Exeter Books

NEW YORK

A Bison Book

First published in USA 1987
 by Exeter Books
Distributed by Bookthrift
Exeter is a trademark of Bookthrift Marketing, Inc.
Bookthrift is a registered trademark of Bookthrift Marketing, Inc.
New York, New York

ISBN 0-671-08926-9

Printed in Hong Kong

Page 1: Voyager 1 looked back at Saturn as it flew beyond the planet in 1980. A few of the spokelike ring features discovered by Voyager appear in the rings as bright patches in this image, at a distance of 3.3 million miles. Saturn's crescent is seen through all but the densest portion of the rings.

Pages 2-3: This spectacular photograph of the southern hemisphere of Jupiter was obtained by the American spacecraft Voyager 2 on 25 June 1979 at a distance of 8 million miles. The spacecraft was rapidly nearing the giant planet, with closest approach on 9 July. Seen in front of the turbulent clouds of the planet is Io, the most volcanically active body in the Solar System, and the innermost of Jupiter's large Galilean moons.

These pages: Two high resolution scans by one of Viking 2's cameras were combined to create this scene, looking northeast from the Utopian Planitia, 48° north of the Martian equator. The largest rock near the center of the picture is about two feet long and one foot high. What appears to be a small channel winds from upper left to the center of the photc. The NASA poetical caption department says, with a flourish, 'On a clear day on Mars, you can see tens of thousands of rocks.'

Designed by Bill Yenne
Edited by Barbara Thrasher and Timothy Jacobs

Picture Credits

All photographs and line drawings were supplied by the National Aeronautics and Space Administration (NASA) unless otherwise noted. All maps were supplied by the Planetary Data Facility of the US Geological Survey unless otherwise noted.

Dean Biggins: 67 (bottom center)
Defense Mapping Agency: 52-53, 54-55, 56-57, 58-59, 60-61, 62-63
Dover Publications: 65 (both), 66 (top right and center)
J D Griggs (US Geological Survey): 45
Hansen Planetarium: 10-11, 38 (center), 178
George Harrison (USFWS): 67 (top center)
Paul Horsted (South Dakota Department of Tourism): 64
Iowa Development Division: 105
Edward La Rue (USFWS): 66 (bottom left)
Wayne Lynch (Parks Canada): 67 (bottom left and center)
© 1986 Max-Planck-Institute für Aeronomie: 170 (right and both upper left), 171 (left)
Ed McCrea (USFWS): 66 (right)
Charles R Smith (USFWS): 66 (upper right)
Smithsonian Institution: 100 (bottom and right)
South Dakota Department of Tourism: 49 (right), 67 (bottom right)
US Fish and Wildlife Service (USFWS): 66 (bottom right)
US National Park Service (USNPS): 67 (upper right)
US Naval Observatory: 171 (right)
John Woods (Parks Canada): 67 (top left)
Wyoming Travel commission: 46
© Bill Yenne: 8, 9, 66 (upper left), 112, 113 (bottom), 110 (top), 126, 128 (top center), 132 (bottom), 133 (bottom), 148 (bottom), 149 (bottom), 160 (bottom), 161 (bottom), 176 (bottom), 177 (bottom)

ACKNOWLEDGMENTS

The author would like to extend his appreciative thanks to Ruth DeJauregui, Gail Rolka and Pam Berkman for their help in collecting and collating the data for the tables in this book; and to Doug Storer for persistently urging me to do this book.

I would also like to thank Dr Mary Kay Hemenway, of the Astronomy Department at the University of Texas, and Bing F Quock, Assistant Chairman of the Morrison Planetarium in San Francisco, for reviewing the original manuscript to ensure the utmost scientific accuracy.

Finally, I'd like to thank the staff of the Planetary Data Facility of the US Geological Survey, specifically Ray Badson, Pat Bridges, Mary Strobell and Jody Swan, for the great deal of help they supplied with cartography and nomenclature.

TABLE OF CONTENTS

INTRODUCTION

The mapping of the Solar System beyond the Earth and its Moon has come a long way since the first Pioneer and Mariner spacecraft began exploring the terrestrial planets in the early 1970s. Solar System cartography, however, is still in much the same place as that of the Earth was in the early sixteenth century. One is reminded that the maps of that era are marked with great empty unknown spaces identified only as *terra incognita*. It should be remembered, however, that it was an immense step, both in the history of cartography and in the history of mankind itself, for mapmakers to be able to conceptualize that there *was* existing *terra* which *was incognita!* Mankind had at last reached the point where it could conceive of the spherical reality of the globe, and this in turn inspired the broadened imaginations and horizons of the brave explorers who struck out to investigate *terra incognita* first hand.

The flights of the Soviet Venera and American Viking spacecraft have provided a great deal of information about Earth's nearest neighbors in the Solar System, and the American Voyager 1 and Voyager 2 spacecraft have provided vastly more information about the great planets of the outer Solar System than was known before 1979 or could ever have been uncovered by Earth-based observers.

The voyages of these spacecraft are the equivalent on a Solar System scale to the sea-borne voyages of global discovery that began with Columbus in 1492. There still exists a great deal of *terra incognita* in the Solar System. This Atlas provides a compendium of what *is* known and what *has* been mapped, as well as a possible point of departure to imagining what further treasures await discovery in the 'backyard' of our own star.

The Universe is so vast that its size is virtually incomprehensible. Even that small portion of it that is visible from Earth is so broad that light waves travelling at the rate of 186,281.7

miles every second will take several billion years to make their way across it. Distances are so great within the Universe that they are measured in *light years,* or the distance that light travels in one year—*5.88 trillion miles.*

The Universe as we know and understand it is filled with untold millions of *galaxies,* which are systematic clusters of *stars* grouped into formations that appear as spirals or discs—technically referred to as *ovoids.* There are, in turn, millions of stars within each of the galaxies spinning and spiralling around the respective galactic centerpoints. There are also *elliptical galaxies,* however, in which stars do not revolve like planets around the Sun, but rather display a motion similar to random kinetic motion.

One ovoid galaxy, known as the Milky Way Galaxy, is home to the star which we call the Sun, and which is the centerpoint of our Solar System. The Milky Way Galaxy is 100,000 light years in diameter, and despite its appearance as a nearly solid mass of starlight, the distances between the individual stars within it are immense. The star nearest our Sun, for example, is Alpha Proxima, part of the Alpha Centauri group, which is four light years or nearly 24 trillion miles away!

The Sun and the entire Solar System are themselves revolving around the center of the Milky Way Galaxy. It seems hard to believe that this delicately balanced Solar System is, in its entirety, hurtling through space at a speed of 630,000 mph.

Our Sun and its Solar System are a relatively tiny part of the Milky Way Galaxy—which is in turn a relatively minute part of the known Universe.

By definition, the Solar System is the Sun and the objects which revolve in orbit around it. The Sun itself accounts for 99 percent of the mass of the Solar System and most of the balance is made up by the nine known *planets* and their *moons* (which

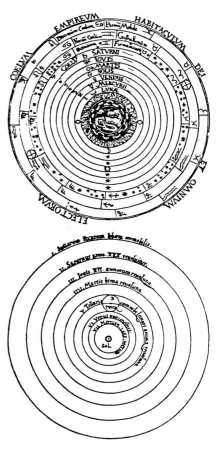

Facing page: Our Solar System, and its position in our infinitely larger galaxy. *Above:* The Copernican and the (*at top*) Ptolemaic conceptions of the universe.

The diagrams below show the known objects that are within 40 Astronomical Units (3.7 billion miles) of the Sun. Among known objects only Pluto at aphelion is farther from the Sun. The top (and largest) section of the diagram shows the relative distances of the Solar System's moons from their respective planets on a *15 million mile scale*. Most of them, however, are less than a million miles away. The scale of the moons is shown in more detail in diagrams on pages 112-113, 132-133, 148-149 and 160-161.

15 Million Miles		The Moons of **Jupiter:**
		• Sinope
		• Pasiphae
14 Million Miles		• Carme
13 Million Miles		
12 Million Miles		• Ananke
11 Million Miles		
10 Million Miles		
9 Million Miles		The Moons of **Saturn:**
		• Phoebe
8 Million Miles		• Elara
		• Lysithea
		• Himalia
7 Million Miles		• Leda
6 Million Miles		
5 Million Miles		
4 Million Miles		The Moons of **Uranus:**
3 Million Miles		Oberon / Titania / Umbriel / Ariel / Miranda
	• Iapetus	
2 Million Miles		1985U1 / 1986U5 / 1986U4 / 1986U1 / 1986U2 / 1986U6 / 1986U3 / 1986U9 / 1986U8 / 1986U7
	• Callisto	

• Ganymede	1980 S1	1980 S25	Hyperion	
The Moon of the **Earth:**	The Moons of **Mars:**	• Amalthea	Europa	1980 S3	1980 S13	Titan
	Deimos	Adrastea	Io	1980 S26	Tethys	Rhea
•	Phobos	Metis	Thebe	1980 S27	Enceladus	1980 S6
				1980 S28	Mimas	Dione

Sun 1 2 3 4 5 6 7 8 9 10 11 12 13 14 15 16 17 18 19 20

Mercury Earth   Jupiter   Saturn   Uranus
Venus   Mars

The Asteroid Belt   *Trojan Asteroids*

1 Ceres
2 Pallas
3 Juno
4 Vesta
10 Hygiea
31 Euphrosyne
2063 Bacchus
1862 Apollo
1,221 Amor
944 Hidalgo
2060 Chiron

Sun 1 2 3 4 5 6 7 8 9 10 11 12 13 14 15 16 17 18 19 20

The center diagram shows the planets them-selves on a *40 astronomical unit scale,* while the bottom diagram shows a selection of Aster-oids on the same scale. The shaded area on the bottom two diagrams is the Asteroid Belt. The Asteroids are indicated by lines rather than dots because they have highly elliptical orbits and hence a big difference between their re-spective aphelions and perihelions. In the case of 1 Ceres this difference is roughly a half AU, but 2060 Chiron swings wildly in an orbit that varies by ten AU, or nearly a billion miles!

	15 Million Miles
	14 Million Miles
	13 Million Miles
	12 Million Miles
	11 Million Miles
	10 Million Miles
	9 Million Miles
	8 Million Miles
	7 Million Miles
	6 Million Miles
	5 Million Miles
	4 Million Miles
	3 Million Miles
	2 Million Miles
	1 Million Miles

The known Moons of **Neptune:**
• Nereid

The known Moon of **Pluto:**
• Charon

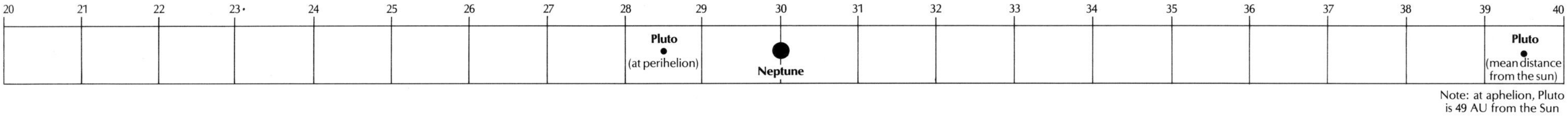

• Triton

20 21 22 23 • 24 25 26 27 28 29 30 31 32 33 34 35 36 37 38 39 40

Pluto
•
(at perihelion)

●
Neptune

Pluto
•
(mean distance
from the sun)

Note: at aphelion, Pluto
is 49 AU from the Sun

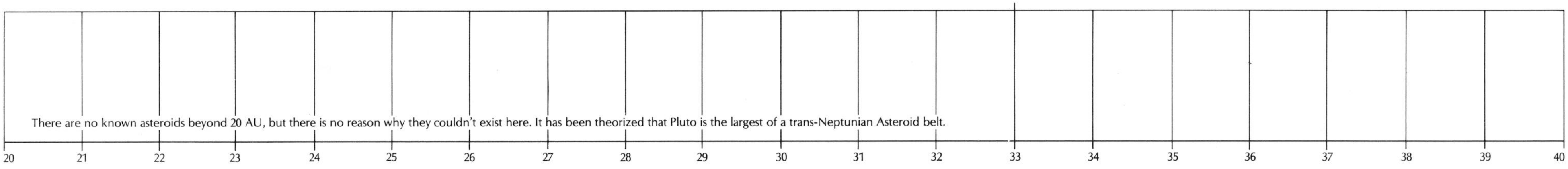

There are no known asteroids beyond 20 AU, but there is no reason why they couldn't exist here. It has been theorized that Pluto is the largest of a trans-Neptunian Asteroid belt.

20 21 22 23 24 25 26 27 28 29 30 31 32 33 34 35 36 37 38 39 40

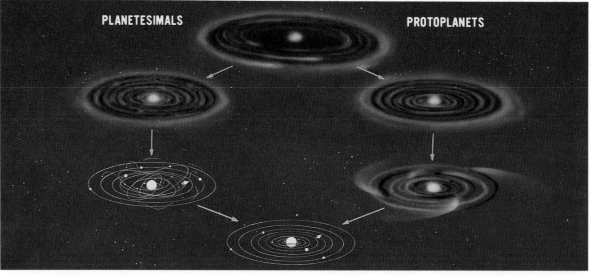

PLANETESIMALS   PROTOPLANETS

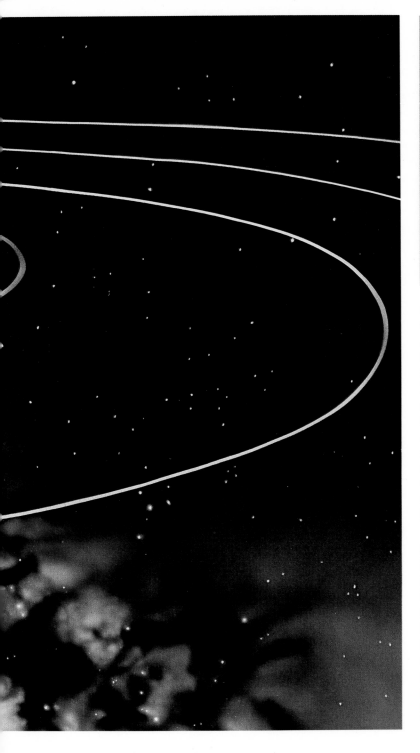

are in orbit around their planets as those planets are in orbit around the Sun). Prior to the 1989 arrival near Neptune of the American observation spacecraft Voyager 2, there were 54 moons positively identified within the Solar System, with more than half of them in orbit around Jupiter and Saturn, the two largest planets. Other objects within the Solar System include *asteroids,* or *minor planets,* which number more than 3000 and exist primarily in a belt between the orbits of Mars and Jupiter. Also present are *meteoroids*—small fragments of rock which exist throughout the Solar System, but which are too small to be seen until they plunge into the Earth's atmosphere leaving their distinctive fiery trails. More spectacular are *comets*—icy objects which appear to take on great, brilliant tails when their extreme-ly elliptical orbits bring them close to the Sun.

The Solar System originated 4.6 billion years ago when *proto-stellar (nebula)* material, a hot swirling cloud of mostly pure hydrogen gas (the simplest of elements), gradually collapsed—succumbing to gravity—and cooled. Gravitational contraction heated this *protostar* and in turn a nuclear fusion reaction was sparked amid the hot, dense gas that condensed at its center, and the Sun was born. The planets were formed from the remaining disc of material still swirling around the Sun, although theories about exactly how this happened disagree. The four largest planets were, and remain, composed largely of hydrogen as well as helium. As such, they and the Sun are relics of the cloud of protostellar nebula material that existed 4.6 billion years ago. The silicate rock, metals, oxygen, nitrogen, carbon and other

Facing page: Our Solar System—outward from the Sun—a typical, highly elliptical comet path; Mercury; Venus; Earth; Mars; Jupiter; Saturn; Uranus; Neptune; and, its unusual orbit represented in yellow (and here shown at the point at which its eccentricity actually brings it within the Neptunian orbit), the outermost known planet, Pluto.

Above: The hypothetical formation of planets from a solar nebula.

Above: The five outermost planets in the Solar System appear in this relationship to the Sun and to each other. On 13 June 1983, the unmanned Pioneer 10 spacecraft was farther from the Sun than any of the Solar System's planets. In this view our Sun is in the middle distance, with the planets shown on the far side of the Solar System from the departing spacecraft. For ease in identification, the planets have been depicted at exaggerated scales and the Sun is obviously much reduced from its relative size. Even with the Sun the size it has been drawn, all of the four terrestrial planets closest to it (Mercury, Venus, Earth and Mars) would be mere pinheads and lost in its glare.

In this illustration the departure of the spacecraft is viewed 'over its shoulder' as Pioneer speeds on its journey. Its radio antenna is trained backward, toward Earth and the Sun. Jupiter is shown in its orbit, with Saturn next out from it. Uranus, with its spin axis tilted at more than 90 degrees from those of the other planets, is the next orbit out. Neptune is next and the orbit line of Pluto crosses the top of this view.

materials found in the other planets and moons are probably relics of the impurities that existed in the original cloud. Heavy elements are believed to originate in other, more massive stars. They are in turn distributed through space when these stars explode as *supernovae.*

The Solar System displays several fundamental regularities in its structure. This seems to indicate that the mechanisms which formed the Solar System were not random but rather were the actions of orderly (if not fully understood) physical processes. The planets are not randomly arranged but rather have regular concentric, near-circular (except for Pluto) orbits. They all revolve in the same direction, and all of the major bodies revolve around the Sun in a relatively flat plane. Using the Earth's orbital plane as the zero degree plane, the orbital planes of all of the other planets tilt no more than 3.39 degrees, except for those of Mercury (7.0 degrees) and Pluto (17.2 degrees).

Theoretically the Solar System extends outward from the Sun to the point (or circular series of points) beyond which the Sun's gravity has no effect. Because theoretical points are hard to measure precisely (and gravity extends indefinitely), we could use the orbital diameter of Pluto, the outermost planet, as the diameter of the Solar System. However, because Pluto's eccentric orbit briefly brings it closer to the Sun than Neptune's more circular orbit, the diameter of what is familiarly known as our Solar System should probably be pegged to the aphelion of Pluto (4.57 billion miles) plus the aphelion of Neptune (2.81 billion miles). Thus the diameter of the Solar System is roughly 7.38 billion miles, .00126 light years, or just short of 11 *light hours.* This formula (or the diameter of Pluto alone) falls short, however, of *truly* defining the limits of the Solar System. At a distance of 5580 billion to 7440 billion miles from the sun there is a spherical cloud containing 2×10^{12} comets with a total mass of seven to eight times that of the Earth. A more massive inner cloud 100 times the mass of the outer cloud stretches from Neptune's orbit to a distance of 930 billion miles. Taking these distances into account would indicate a Solar System diameter on the order of 2.5 light years.

Distances *within* the Solar System, while infinitesimal on the galactic scale, are quite large, so astronomers measure intra-Solar System distances in *Astronomical Units* (AU), which are equivalent to the distance between the Earth and the Sun—93 million miles or 8.3 *light minutes.*

If a solar system such as ours could form out of a cloud of protostellar nebula material, it is certainly reasonable to assume that planets have formed around others of the billions of stars that exist in the Milky Way Galaxy and beyond. Since our Solar System is governed by—and was probably formed by—orderly physical processes, it is more probable to assume that other Solar Systems do exist than to assume that they do *not.* Disks of material similar to the disk that was theorized to have formed our Solar System have, in fact, already been observed around other stars, such as Beta Pictoris, Fomalhaut and Vega. In December 1984, Dr Donald McCarthy and Dr Frank Low of the University of Arizona and Dr Ronald Probst of the US National Optical Astronomy Observatory at Tucson detected a nonstellar object in orbit around the star Van Biesbroek 8. It was heavier and more massive than Jupiter and had a surface temperature equivalent to molten lava. Officially classified as a brown dwarf star, the object has not been seen since its original discovery. If relocated, however, it could ultimately be confirmed as the first planet to be identified beyond our Solar System.

Our own Solar System can generally be organized into six parts or zones. Moving outward from the Sun they are:

1. The *terrestrial,* or solid surfaced, planets (Mercury, Venus, Earth and Mars) with their total of only three moons which span the first 1.6 AU from the Sun.

2. The Asteroid Belt which spans the 3.8 AU distance from the orbit of Mars to the orbit of Jupiter. Most, but not all, known asteroids are to be found within this belt.

3. Beginning 5.4 AU from the Sun, and spanning a distance of 24.8 AU, the third and widest zone includes the four largest planets (Jupiter, Saturn, Uranus and Neptune) along with their 50 moons. These planets are identified as *gas giants* because of their composition and because they are much larger than any other body in any other zone.

4. The final zone of the familiar Solar System contains the planet Pluto and its single moon, which orbit in an elliptical path that ranges between 29.5 AU to 49.2 AU from the Sun.

5. The inner of two clouds of comets that extends from 30 AU to 10,000 AU.

6. The second of two clouds of comets that extends from 60,000 AU to 80,000 AU.

Each of the two major groups of planets contains *four* major bodies that are as similar to one another as they are dissimilar to the planets in the other group. The terrestrial planets range in diameter between three and eight thousand miles while the gas giants have a size range almost exactly 10 times greater. The terrestrial planets all have solid silicate rock crusts with interiors that are (or once *were*) molten, while the gas giants are balls of hydrogen and helium and bear a closer resemblance to the composition of the Sun than they do to the composition of the terrestrial planets. The four terrestrial planets have just three moons between them, while the four gas giants have at least 50 among *them*.

The composition of the moons of the outer planets is, however, intriguingly similar to that of the terrestrial planets themselves. Almost all of them are composed of water, ice and silicate rock. Because of their hydrogen/helium composition, the gas giants are known to be closely related to the Sun, so their moon systems could be looked upon *almost* as solar systems within a solar system. Jupiter, for example, is known to have originally been a star whose origin was much like that of the Sun, but which was not massive enough to have undergone self-sustaining fusion reactions.

The bodies in the Solar System can be classified in another important way. Out of the more than 3000 planets, moons and asteroids in the Solar System, only *eight* are known to have atmospheres consisting of more than barely detectable traces of gases near their surfaces. These include the four gas giants, of course, which could be described almost as being *all* atmosphere. The others are the terrestrial planets Venus, Earth and Mars as well as Saturn's moon Titan. Of these four, Venus and Titan have atmospheres that are so thick that their solid surfaces are completely obscured by clouds.

The objects in the Solar System can also be classified by their surface type. Again, of course, the gas giants with their gaseous 'surfaces' are in a class by themselves. Another class would be those with silicate rock surfaces that have been marked primarily by meteorite impact craters. This class would include Mercury, the Earth's Moon, the Martian moons and all the asteroids. A third class, the so-called 'dirty snowballs,' are composed mostly of silicate rock and water ice marked by meteorite impact craters and some inherent geologic activity. This class would include nearly all the moons of the outer Solar System's four gas giants.

Five of the major bodies in the Solar System have surfaces that could truly be classed as unique. Erosion by liquid water has played a significant role in forming the surface features of two planets—Earth and Mars—but while it still plays that part on Earth, liquid water has mysteriously vanished from the surface of Mars. Huge oceans of liquid water and bits of water ice cover 70 percent of the Earth, while a similar percentage of Titan's surface is probably covered with frigid seas of liquid methane

The five largest objects in the Solar System—the Sun itself, Jupiter, Saturn, Uranus and Neptune—are essentially gaseous. All the smaller bodies have solid surfaces and of these, ten of the eleven largest are pictured *at left*. Not shown is Neptune's moon Triton that rivals Ganymede as the Solar System's largest moon. In the next tier below these in size are four Saturnian moons, three Uranian moons, a Jovian moon, the planet Pluto and Ceres, the largest asteroid.

filled with methane icebergs. Every solid surfaced body in the Solar System has been, and still is, susceptible to marring by the impact craters of meteorites, but Venus, Titan and the Earth have atmospheres that are so dense that only in rare instances would a meteorite be large enough to not burn up in those atmospheres. Only a handful of bodies in the Solar System have ever had active volcanos and only two, the Earth and Jupiter's Io, are known to still have them. The most volcanically active object in the Solar System, Io, is home to almost continuous eruptions by sulfurous volcanos, from which liquid sulfur 'lava' constantly resurfaces the planet. This has in turn resulted in there being no identifiable meteorite impact craters on Io.

The Solar System is an amazing place, full of intriguing similarities and inexplicable peculiarities. It is amazing in both its orderliness and its diversity. It interests us and confuses us, for as much as we are able to learn, our new knowledge merely serves to inspire new questions.

THE SUN

The central body of our Solar System, the Sun is an average-sized star of the *yellow dwarf* variety that formed roughly 4.6 billion years ago at the center of the enormous swirling gas cloud that became the Solar System. The concentration of pressure at the center of this swirling cloud of (mostly) hydrogen triggered a nuclear fusion reaction and the star we know as the Sun was born. In this fusion reaction, typical of all stars, four nuclei of hydrogen atoms (the simplest of the elements) fused to form a single helium (the second simplest element, having two protons) nuclei. The resulting reaction released a tremendous amount of energy.

It is theorized that in about five billion years, as its hydrogen becomes depleted, the Sun will expand from its present status of yellow dwarf star to become a *red giant,* with a diameter greater than the orbit of Venus. According to this generally accepted theory, the Sun will then collapse back to a *white dwarf* type star (smaller than its present size), and gradually become a burnt ember which would then be ironically referred to as a *black dwarf.*

Though it is not considered to be a particularly large star, the Sun is by far the largest body within the Solar System, containing 99 percent of the matter of the system, thus providing the gravitational force that literally *defines* the Solar System and controls the orbital paths of the other bodies within it. The Sun is also the source of most of the heat in the Solar System and thus it provides the warmth that makes life possible on at least one of the bodies in the Solar System. The temperature at the core of the Sun is estimated at roughly 20 million degrees centigrade, and the Sun's surface temperature averages 11,000 degrees Fahrenheit or approximately 6000 degrees Kelvin.

The energy radiated from the Sun is called *solar radiation,* which (as measured in wavelengths from the longest to the shortest) can be simplified as including: (a) radio waves, (b) microwaves, (c) infrared radiation (perceived on earth as heat), (d) the visible light spectrum, (e) ultraviolet radiation, (f) x-rays and (g) gamma rays. So powerful is solar radiation that its ultraviolet wavelengths can burn (or tan) human skin on Earth, and direct light from the visible spectrum can do permanent damage to the human eye.

Like the other major bodies in the Solar System, the Sun rotates on its axis. However, its equatorial region rotates once every 27.275 days while its polar regions have a slower rotational period of 34 days.

THE PHOTOSPHERE

The Sun, being a gaseous sphere, has no solid surface, nor could any molecular solid exist at such incredible temperatures. The Sun does, however, have a nearly opaque surface—a sea of gaseous firestorms known as the *photosphere.*

The firestorms that comprise the photosphere are roughly 600 miles in diameter and appear as granules in the vastness of the Sun. Their apparent opacity is due to the presence of negative hydrogen ions. During the approximate eight-minute lifespan of the granule, hot gas rises out of the center, pushing cooler gases aside and into the narrow darker and cooler spaces between the

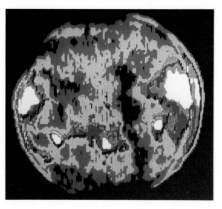

Facing page: A huge solar flare can be seen in this enlarged spectroheliogram, obtained during the United States Skylab 3 mission in 1974. In this photograph, helium erupting from the Sun has remained a coherent mass to an altitude of 500,000 miles.

The relative intensity of solar regions is color coded in this Orbiting Solar Observatory television display (*above*). White represents the greatest intensity, followed in descending order by yellow, red and blue. The dark regions at the poles, extending down across the face of the Sun, are coronal holes.

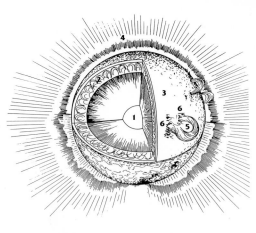

Above is a diagram of the illustration of the Sun (shown on the facing page) showing (1) the core; (2) the convection zone; (3) the photosphere; (4) the chromosphere; (5) a solar prominence; and (6) sunspots.

granules. Amid the typical granules, there are 'supergranules' with diameters of up to 18,000 miles and lifespans of up to 24 hours.

Other 'surface features' on the photosphere are *solar flares* and *sunspots*. Solar flares are violent surface eruptions that explode from the photosphere with the energy of 10 million hydrogen bombs, sending forth a stream of solar radiation that can disrupt radio signals on the Earth.

Solar flares were first observed in 1859 by the English astronomer Richard Carrington. They have also been observed to a greater or lesser degree on other stars. The frequency of solar flares can range from several in a single Earth day during periods when the Sun is active, to fewer than one per Earth week during the periods which astronomers describe as 'quiet.' The energy for individual flares may take several hours or even days to build up, but the actual flare happens in a matter of minutes when the energy is released. The resulting shockwaves travel outward across the photosphere and up into the chromosphere and corona for hundreds of thousands of miles at speeds on the order of three million mph.

It is not known what triggers solar flares, but magnetic energy almost certainly plays a major role. The study of solar flares is important because of the effect on Earth of the radiation and particles released during solar flares, not to mention the potential negative effect on spacecraft and astronauts beyond the Earth's atmosphere. The charged particles released in the flares are attracted by the Earth's magnetic field and spiral in at the north and south magnetic poles, causing the Aurora Borealis in the Earth's atmosphere.

Sunspots are dark regions on the surface of the photosphere which are cooler than the surrounding areas. Like the solar flares, sunspots occur with less frequency during the sun's quiet periods. During such periods there may be *no* observable sunspots, while during active periods there may be more than a hundred on the photosphere at one time.

Sunspots were first observed by the Chinese 2000 years ago, and in the seventeenth century, the great Italian astronomer Galileo Galilei (1564-1642) conducted a systematic study of them. His observations of the motion of sunspots across the solar surface led to his discovery of the rotation of the Sun. The frequency of sunspot activity has been recorded as occurring in an 11 year cycle that seems to have an effect on the weather on Earth. The period from 1645 to 1715 was, for example, an era of a very quiet Sun and for *seven years* during this time *no* sunspots were observed. These years also corresponded with the height of the cold spell in the Earth's northern hemisphere that is referred to as the 'little ice age.'

Sunspots vary in size and shape and may be as large as 40,000 miles across. They are composed of a 'penumbra' with a darker 'umbra' in the center (which constitutes about a quarter of the sunspot's area). Sunspots increase to their full size in about a week to 10 days, but in turn take nearly two weeks to decay. Sunspots usually occur in groups, and a large group may have a life span of several weeks.

It is not known what causes sunspots, but the standard theory has it that a powerful magnetic field temporarily restricts the flow of the hottest gasses to that particular part of the photosphere; sunspots seem to appear at places where magnetic field lines have become twisted and rise above the photosphere.

THE CHROMOSPHERE

Above the photosphere there is a thinner, more visually transparent layer known as the *chromosphere* (literally 'color sphere'). This layer is of indeterminate depth, but it is roughly 6000 miles thick. The most common feature within the chromosphere are the *spicules,* long thin fingers of luminous gas which appear like a vast field of blades of fiery grass growing up into the chromosphere from the photosphere. They are observed to rise to the upper reaches of the chromosphere (about 6000 miles

Data for The Sun

Diameter: 870,331.25 miles (1,392,530 km)
Mass: 4.3959×10^{30} lb (1.9891×10^{30} kg)
Rotational period
  of equator: 26.8 Earth days
  at latitude 30°: 28.2 Earth days
  at latitude 60°: 30.8 Earth days
  at latitude 75°: 31.8 Earth days
Surface temperature: 10,430° F
Interior temperature: 26,999,540° F
Major photospheric components:

Hydrogen (73.46%)	Neon (.12%)
Helium (24.85%)	Nitrogen (.09%)
Oxygen (.77%)	Silicon (.07%),
Carbon (.29%)	Magnesium (.05%)
Iron (.16%)	Sulphur (.04%)

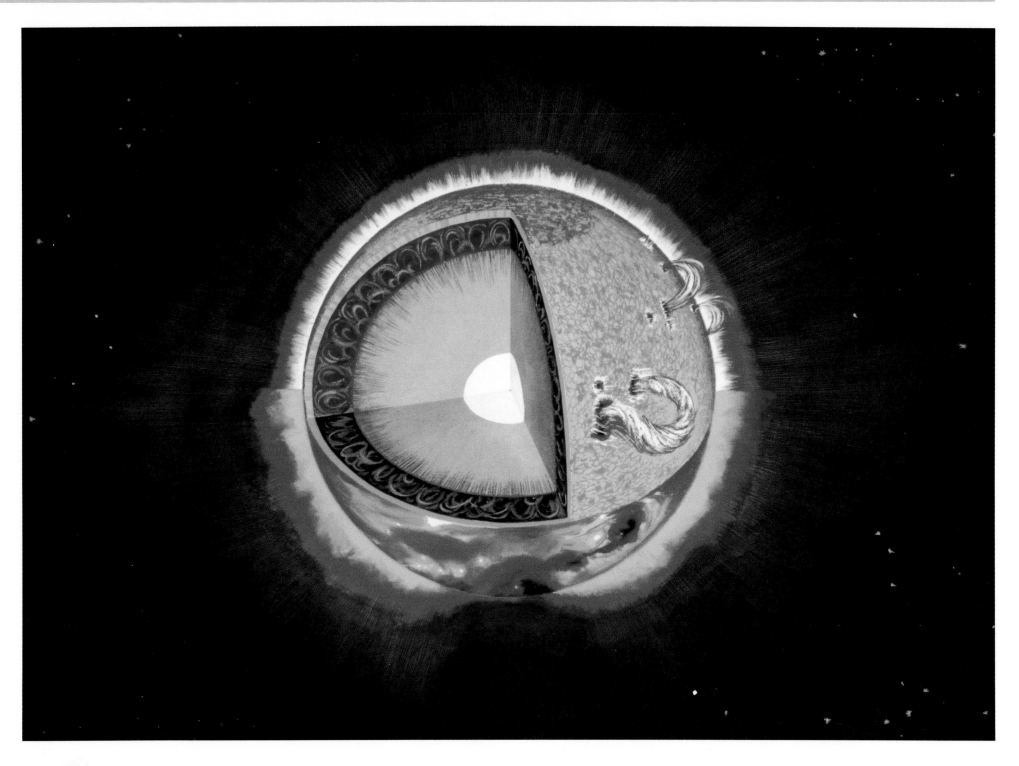

Successful Solar Spacecraft Missions

Spacecraft	Country of Origin	Launch	Results	Spacecraft	Country of Origin	Launch	Results
Pioneer 4	USA	3 Mar 59	Studied Earth's magnetic field and solar flares (solar orbit)	OGO 4	USA	28 Jul 67	Studied Earth-Sun relationships, atmospheric ionization, Earth's magnetic field and aurorae
Vanguard 3	USA	18 Sep 59	Studied solar X-radiation				
Pioneer 5	USA	11 Mar 60	Studied solar flares and Solar Wind (solar orbit)	OSO 4	USA	18 Oct 67	Studied the Sun's very shortwave ultraviolet radiation and other general activity
OSO 1 Orbiting Solar Observatory	USA	7 Mar 62	Studied solar flares, Earth orbit of 901 miles, sent back data on 75 flares; ended 6 Aug 63	Pioneer 8	USA	13 Dec 67	Studied solar radiation (solar orbit)
Cosmos 3	USSR	24 Apr 62	Studied solar and cosmic radiation, density of Earth's upper atmosphere	OGO 5	USA	4 Mar 67	Studied general solar activity and Earth's magnetic field
				Explorer 37 Solrad	USA	5 Mar 68	Studied solar activity and radiation (solar orbit)
Cosmos 7	USSR	28 Jul 62	Studied/monitored solar flares during manned flights of Vostok 3 and 4	Cosmos 230	USSR	6 Jul 68	Studied general solar activities
				Pioneer 9	USA	8 Nov 68	Studied general solar activities (solar orbit)
Explorer 18 Interplanetary Monitoring Platform 1	USA	26 Nov 63	Studied/monitored solar flares during manned Apollo and Skylab missions	HEOS 1	USA	5 Dec 68	Studied interplanetary solar particles (solar orbit); with HEOS 2 covered 7 years out of the 11-year solar cycle
OGO 1 Orbiting Geophysical Observatory 1	USA	4 Sep 64	Studied Earth-Sun relationships and the Sun's effect on the Earth's magnetic field	Cosmos 262	USSR	26 Dec 68	Studied solar ultraviolet and X-radiation (Earth orbit)
OSO 2	USA	3 Feb 65	Studied solar activity, measurements of ultraviolet, X-ray and gamma radiation	OSO 5	USA	22 Jan 69	Studied solar activity and flares
				OGO 6	USA	5 Jun 69	Studied solar influence on Earth's ionosphere and aurora
OGO 2	USA	14 Oct 65	Studied solar ultraviolet and X-radiation and the Sun's effect on Earth's magnetic field	OSO 6	USA	9 Aug 69	Studied solar flares, the corona and general solar activity
Explorer 30 Solrad	USA	18 Nov 65	Studied solar radiation in the IYQS (International Year of the Quiet Sun)	Intercosmos 1	USSR/ International	14 Oct 69	Studied solar ultraviolet and X-radiation
				Azur	West Germany	8 Nov 69	Studied flux of solar particles and their effects; solar sychronous orbit
Pioneer 6	USA	16 Dec 65	Studied solar atmosphere; with Pioneer 7, studied data from a strip of the solar surface extending halfway around the Sun (solar orbit)	Intercosmos 4	USSR/ International	14 Oct 70	Studied Earth's magnetosphere and solar ultraviolet and X-radiation
				Explorer 44 Solrad	USA	8 Jul 71	Studied solar radiation
OGO 3	USA	7 Jun 66	Studied Earth-Sun relationships, the Solar Wind, cosmic radiation and the geocorona	Shinsei SSI	Japan	28 Sep 71	Studied solar and cosmic rays
				OSO 7	USA	29 Sep 71	Studied solar flares, the corona and general solar activity
Pioneer 7	USA	17 Aug 66	Studied same program as Pioneer 6 (solar orbit)	Intercosmos 5	USSR/ International	2 Dec 71	Studied general solar activity with Soviet and Czech equipment
OSO 3	USA	8 Mar 67	Studied general solar activity with emphasis on flares	HEOS 2	USA	31 Jan 62	Studied high energy solar particles (Earth orbit originally)
Cosmos 166	USSR	16 Jun 67	Studied solar X-radiation				

above the photosphere), and then drop back in the space of about 10 minutes.

Fibrils are horizontal wisps of gas that drift through the chromosphere. They are of about the same extent as the spicules, but have about twice the duration.

Prominences are gigantic luminous plumes of gas that appear like tongues of flame. They leap from the photosphere into, and beyond, the chromosphere, sometimes reaching altitudes of 100,000 miles. Aside from the less frequent solar flares, prominences are the most spectacular of solar phenomena.

THE CORONA

Beyond the chromosphere is the *corona,* a vast field of hydrogen particles that extends for millions of miles into space. The corona is so sparse that it is not visible against the glare of the Sun except during a total solar eclipse—when the Moon passes between the Earth and the Sun, blotting out the photosphere. During periods of quiet Sun, the corona is more or less confined to the solar equatorial regions with *coronal holes* being present in the polar regions. During periods of more activity, the corona

Spacecraft	Country of Origin	Launch	Results	Spacecraft	Country of Origin	Launch	Results
Cosmos 484	USSR	6 Apr 72	Studied solar and cosmic radiation	Aryabhata	India	19 Apr 75	Studied solar neutrons and gamma radiation
Prognoz 1	USSR	14 Apr 72	Studied Solar Wind and X-radiation; Earth's magnetosphere	OSO 8	USA	21 Jun 75	Studied solar ultraviolet and cosmic X-radiation
Prognoz 2	USSR	29 Jun 72	Studied same program as Prognoz 1	Prognoz 4	USSR	22 Dec 75	Studied solar radiation and geomagnetic field associated with IMS (International Magnospheric Study)
Intercosmos 7	USSR/ International	30 Jun 72	Studied solar shortwave radiation; controlled by a Russian/Czech/East German team	Helios 2	West Germany	15 Jan 76	Studied same program as Helios 1
Aeros 1	West Germany	16 Dec 72	Studied ultraviolet radiation; Vandenburg AFB launch (Earth orbit)	Intercosmos 16	USSR/ International	27 Jul 76	Carried Russian and Swedish solar equipment
Prognoz 3	USSR	15 Feb 73	Studied solar flares, X-radiation and gamma rays	Prognoz 5	USSR	25 Nov 76	Studied Solar Wind, gamma and X-radiation associated with IMS (See Prognoz 4)
Intercosmos 9	USSR/ International	19 Apr 73	Studied solar radiation and activity	Prognoz 6	USSR	22 Sep 77	Studied effects of solar gamma and X-radiation on Earth's magnetic field; ultraviolet, gamma and X-radiation from the galaxy
Skylab	USA	14 May 73	Manned space station carrying 3 successive crews, conducted a wide range of solar science experiments	Prognoz 7	USSR	30 Oct 78	Studied solar ultraviolet and gamma radiation of Earth's magnetosphere
Taiyo (MS-T2)	Japan	16 Feb 74	Studied solar ultraviolet and soft X-rays	Solar Maximum Mission (SMM)	USA	14 Feb 80	Studied the Sun in all aspects near the maximum of the solar cycle
Intercosmos 11	USSR/ International	17 May 74	Studied solar ultraviolet and X-radiation	Solar Mesosphere Explorer (SME)	USA	6 Oct 81	Studied Solar effects on atmospheric ozone
Explorer 52	USA	3 Jun 74	Studied general activity and Solar Wind				
Aeros 2	West Germany	16 Jul 74	Studied general solar activities				
Helios 1	West Germany	10 Dec 74	Studied the Solar Wind and solar surface at close range; passed the Sun 15 Mar 75				

Planned projects include Ulysses (USA/European Space Agency) (1992) and HESP-1 (Japan) (1992).

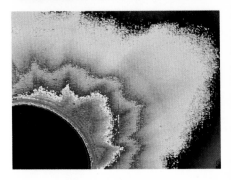

At top and above: In these striking computer-enhanced views of the solar corona, the colors represent densities of the corona and decline from purple (the most dense) to yellow (the least dense). The view at *above right* uses the same color coding to investigate gas densities in a solar flare.

Facing page: Photographed on 19 December 1973, this is one of the most spectacular solar flares ever recorded, spanning more than 367,000 miles across the Sun's surface. The solar poles are distinguishable in this photo by their comparatively uniform dark coloration. Note also the pervasive dark granules, or short-lived firestorms, which average 600 miles in diameter.

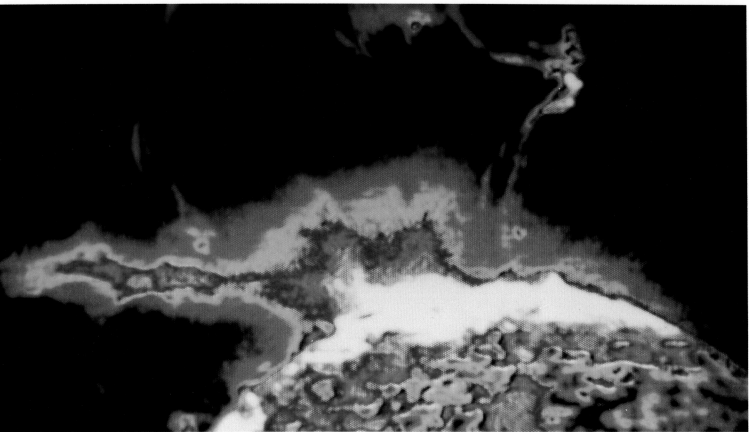

is evenly distributed around the Sun, including the polar regions, but appears most prominent near the regions of the most sunspot activity.

The corona is mysteriously hotter than the photosphere, despite the second law of thermodynamics which holds that heat cannot be conducted from the cooler to the warmer. The mystery involves the process by which the corona is heated. The dynamics of solar magnetic fields and acoustic energy are suggested as possible answers. Despite its high temperature, the corona's very low density means that it radiates relatively little energy.

Triggered by large prominences or by solar flares, *coronal transients* are blast waves, giant loops of corona material released, at speeds of more than a million mph, into the Solar System. Coronal transients are carried by the solar wind to distances beyond the Earth's orbit. Coronal transients have been observed as having 10 times the energy as the flares which trigger them.

THE SOLAR WIND

Blowing outward from the Sun and its corona is a constant stream of hot ionized subatomic particulate plasma known as the *solar wind*. A constant phenomena, the solar wind gusts from 450,000 mph to 2 million mph, and blows into the distant reaches of the Solar System.

The solar wind spirals out from the Sun, rotating with the Sun until it reaches a distance of approximately 100 million miles (roughly 1 AU). From that point it travels outward with less interference from the Sun's magnetic field.

Approximately 3000 tons of subatomic particles are blown outward from the Sun in the solar wind every hour. To give some idea of the scale of the Sun, it would, at this rate, take 200,000 billion years for the Sun's entire mass to be dissipated by the solar wind.

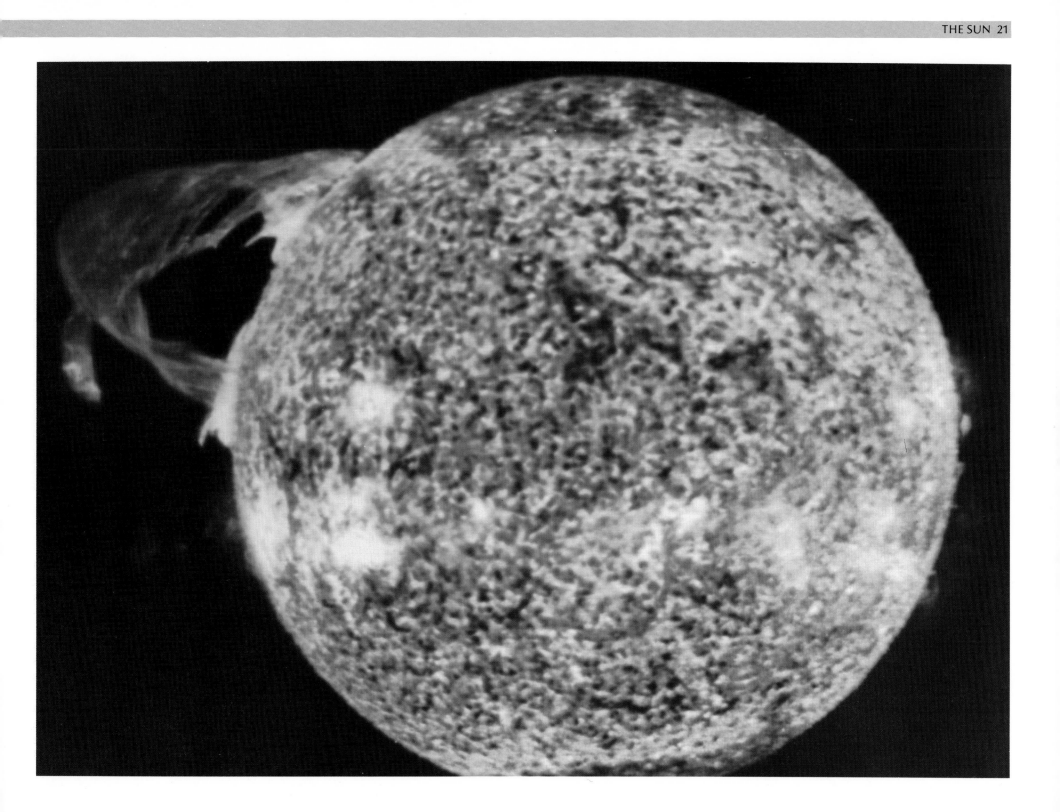

MERCURY

Mercury is the closest planet to the Sun and the second smallest of the nine planets. It has been observed from Earth since prehistoric times, but because of its size it is fainter than Venus, Mars, Jupiter and Saturn—the other planets visible to the naked eye.

Because of its proximity to the Sun, it is always observed within 27 degrees of the Sun in the east before sunrise or in the west after sunset. It has a *sidereal period* of just three months, the shortest of any planet. As such it has the appearance from Earth of moving faster than the others, a characteristic which led the Greeks to name it Hermes, after the messengers of the gods. The Romans in turn called the planet Mercury after their own dieties' wing-footed messenger.

In the eighteenth century Johann Hieronymus Schroeter (1745-1816) became the first astronomer to record his observations of Mercury's surface detail, but his drawings, like those of Giovanni Schiaparelli (1835-1910) more than a century later, were ill-defined and turned out to be inaccurate. The great American astronomer Percival Lowell (1855-1916) reported that he had observed streaks on Mercury's surface similar to those that he and Schiaparelli had both observed on Mars. Shiaparelli had called these Martian features *canali* (channels) and Lowell decided they were *canals,* built by intelligent life. Both astronomers agreed, however, that the streaks on Mercury's surface were of natural origin.

In 1929 Eugenios Antoniadi (1870-1944), a Greek-born astronomer working in France, completed a chart of Mercury's surface that stood for nearly half a century as the accepted map of the planet. Using one of the world's best telescopes, he identified a number of major surface features and proved that the streaks seen by Schiaparelli and Lowell were optical illusions. He agreed with the earlier astronomers, however, on the idea that Mercury's rotational and sidereal periods were identical, thus that the planet always had the same hemisphere turned toward the Sun. It was not until 1962 that this was shown to be untrue.

The major milestone in the observation of Mercury came in March 1974 when the American spacecraft, Mariner 10, began a series of three flybys at a distance of about 12,000 miles, in which it was able to photograph, in great detail, objects as small as 325 feet across.

The photos returned by Mariner 10 revealed a planet whose surface features might easily be mistaken for those on the Earth's Moon. Like the Lunar surface, that of Mercury is pocked by thousands of craters. With the exception of the relatively smooth Caloris Basin, Mercury's surface is characterized almost exclusively by craters, overlapping craters and craters within craters. The Lunar surface, by contrast, has more large open areas known as *maria* or seas. The Caloris Basin, which is itself pocked by hundreds of relatively smaller craters, is the only major open plain comparable to the Moon's maria. Unlike the Lunar seas, which are ancient lava flows, it is believed that the Caloris Basin was created by a massive ancient impact, as is indicated by the presence of mountains and ridges around its periphery, possibly caused by seismic waves. Other ridges and escarpments are to be seen on the surface, and are possibly due to the expansion and contraction of Mercury's core as it cooled and shrank. Some of the cliffs produced by this effect rise as much as 6300 feet above the adjacent valley floors. There is

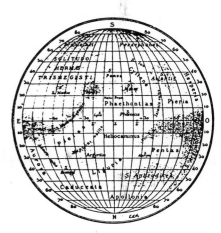

Facing page: Mercury's southern hemisphere glows bleakly in the intense rays of the Sun. Note this planet's heavily cratered surface. *Above:* Eugenios Antoniadi's 1929 map of Mercury.

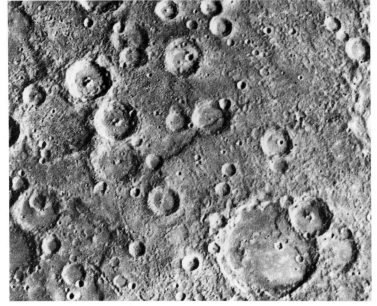

Above: This Mariner 10 view of Mercury's northern region shows a prominent east-facing scarp extending from the planetary 'limb' near the middle of the photo southward for hundreds of miles. The 'tear' near the top of this picture is an incompletion in the computer data used by Earth-based sensors to reconstitute this image.

At right: This scarp or cliff, which towers 7500 feet above the surrounding area, is part of a large system of Mercurian faults which extend for hundreds of miles.

Facing page: Mariner 10's television cameras obtained this view of Mercury on 29 March 1974.

some evidence of ancient volcanic activity on Mercury, but less than that on the Moon.

Because of Mercury's overall density, its core is thought to be largely (70 percent) composed of iron, with the surface crust being silica rock like that of Earth or Lunar surfaces. Due perhaps to its slow rotation, Mercury has a relatively weak magnetic field despite its being composed mostly of iron.

Unlike the other three inner terrestrial planets, Mercury has virtually no atmosphere. Faint traces of gaseous helium form 98 percent of Mercury's 'atmosphere' with the remainder being composed mostly of hydrogen. Minute traces of argon and neon are also present. The helium was probably captured from the Sun because any gases emanating from the interior of the planet would have long ago dissipated into space.

Because of its virtually non-existent atmosphere, Mercury's surface temperatures vary widely. The mid-day temperature on the side facing the Sun can be as hot as 620 degrees Fahrenheit, while at night temperatures can plummet to -346 degrees Fahrenheit because there is no atmosphere to hold the heat.

Data for Mercury

Diameter: 3031 miles (4878 km)
Distance from Sun: 43,309,572 miles (69,700,000 km) at aphelion
28,520,937 miles (45,000,000 km) at perihelion
Mass: 1.501×10^{32} lb (3.302×10^{32} kg)
Rotational period (Mercurian day): 58.65 Earth days
Sidereal period (Mercurian year): 87.97 Earth days
Eccentricity: 0.206
Inclination of rotational axis: 0°
Inclination to ecliptic plane (Earth = 0): 7°
Albedo (100% reflection of light = 1): .06
Mean surface temperature: 407° F
Maximum surface temperature: 620° F
Minimum surface temperature: 194° F (-346° F on dark side)
Largest known surface feature: Caloris Basin
838.9 miles in diameter
(1350 km)
Major atmospheric component: Trace amounts of Helium
Other atmospheric components: Hydrogen, Argon, Neon

Above is a Mariner 10-generated photomosaic of Mercury showing (*inset*) the bright crater Kuiper—which is 25 miles in diameter and was named in honor of the late Dr Gerard Kuiper—as it is positioned on the rim of a larger crater.

The largest structural feature yet discovered on Mercury is seen in the left half of the photomosaic *at right:* it is the Caloris Basin, 800 miles in diameter and bounded by mountains as high as 1.2 miles.

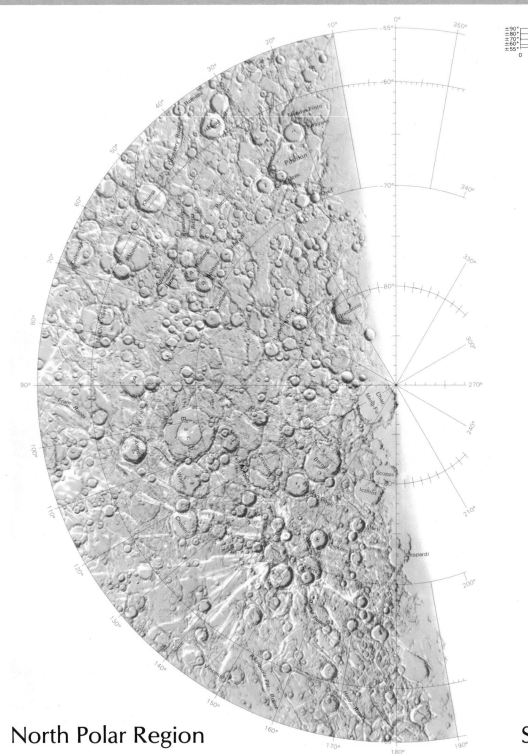

North Polar Region

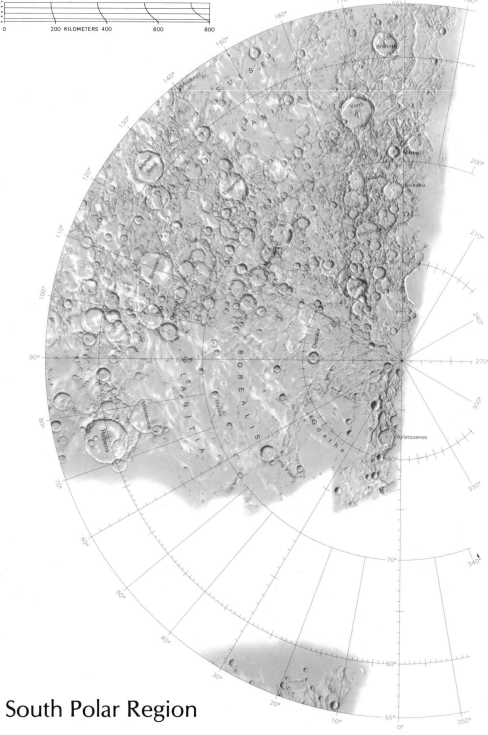

South Polar Region

±90°
±80°
±70°
±60°
±55°

0   200 KILOMETERS 400   600   800

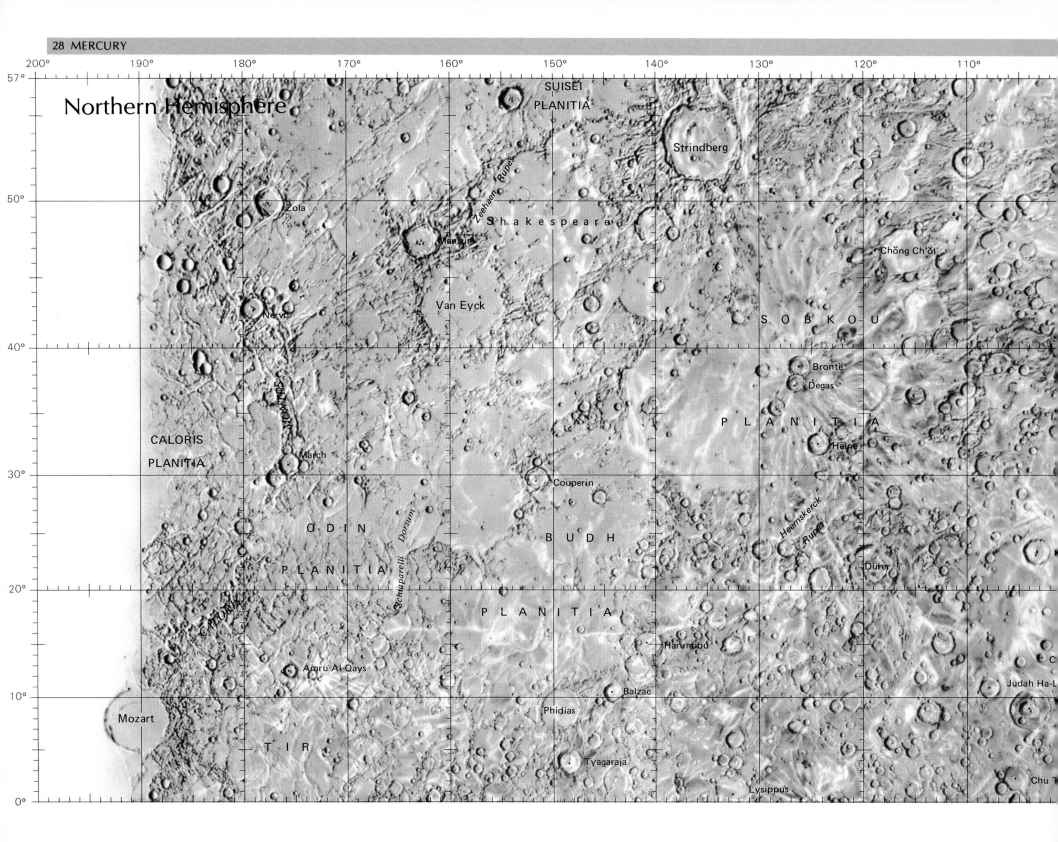

Northern Hemisphere

SUISEI
PLANITIA

Strindberg

Zola

Zeehaen Rupes

Shakespeare

Mansur

Chŏng Chŏl

Van Eyck

Nervo

S O B K O U

Brontë
Degas

CALORIS

PLANITIA

P L A N I T I A

March

Heine

Couperin

O D I N

Dorsum

B U D H

Heemskerck
Rupes

Schiaparelli

Dürer

P L A N I T I A

P L A N I T I A

CALORIS

Harunobu

Amru Al-Qays

Judah Ha-L

Balzac

Mozart

Phidias

Chu T

T I R

Tyagaraja

Lysippus

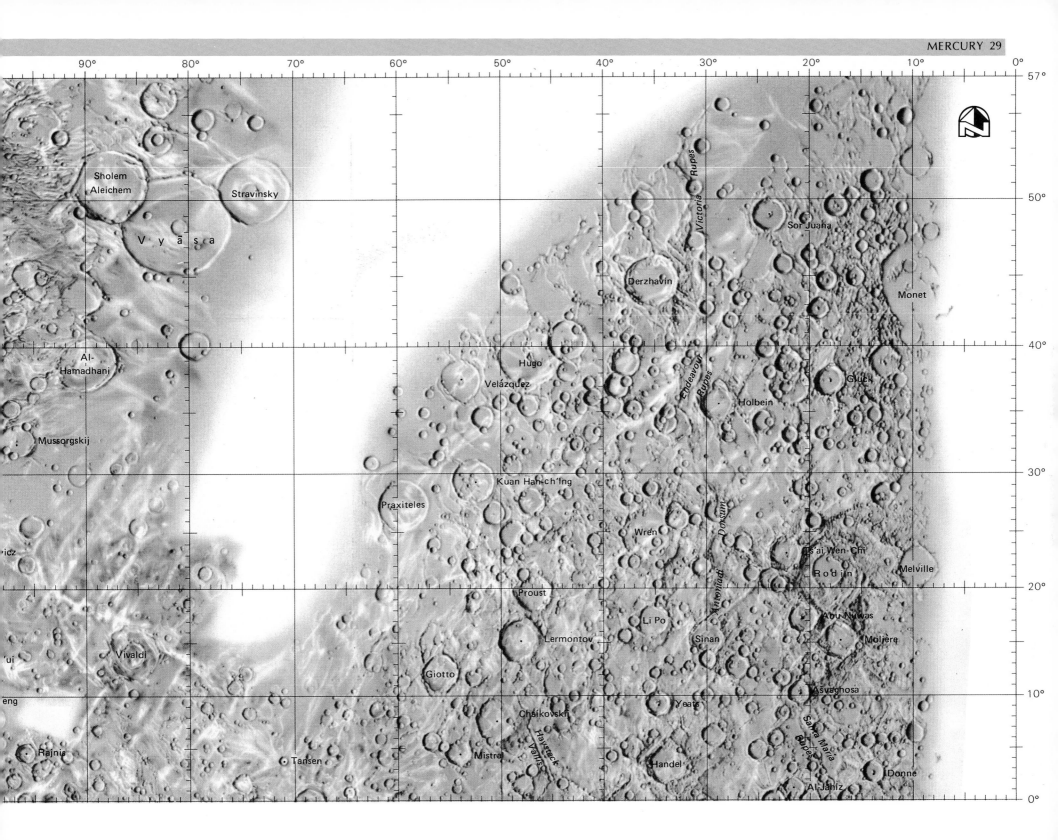

Sholem
Aleichem
Stravinsky
V y ā s a
Al-
Hamadhani
Mussorgskij
icz
Vivaldi
'ui
eng
Rajnis
Tansen
Praxiteles
Kuan Han-ch'ing
Hugo
Velázquez
Derzhavin
Victoria Rupes
Sor Juana
Monet
Endeavour Rupes
Holbein
Gluck
Dorsum
Antoniadi
Wren
Ts'ai Wen-Chi
Rodin
Melville
Proust
Li Po
Abu Nuwas
Lermontov
Sinan
Molière
Giotto
Asvaghosa
Chaikovskij
Yeats
Santa Maria Rupes
Handel
Donne
Mistral
Al-Jahiz
Haystack Vallis

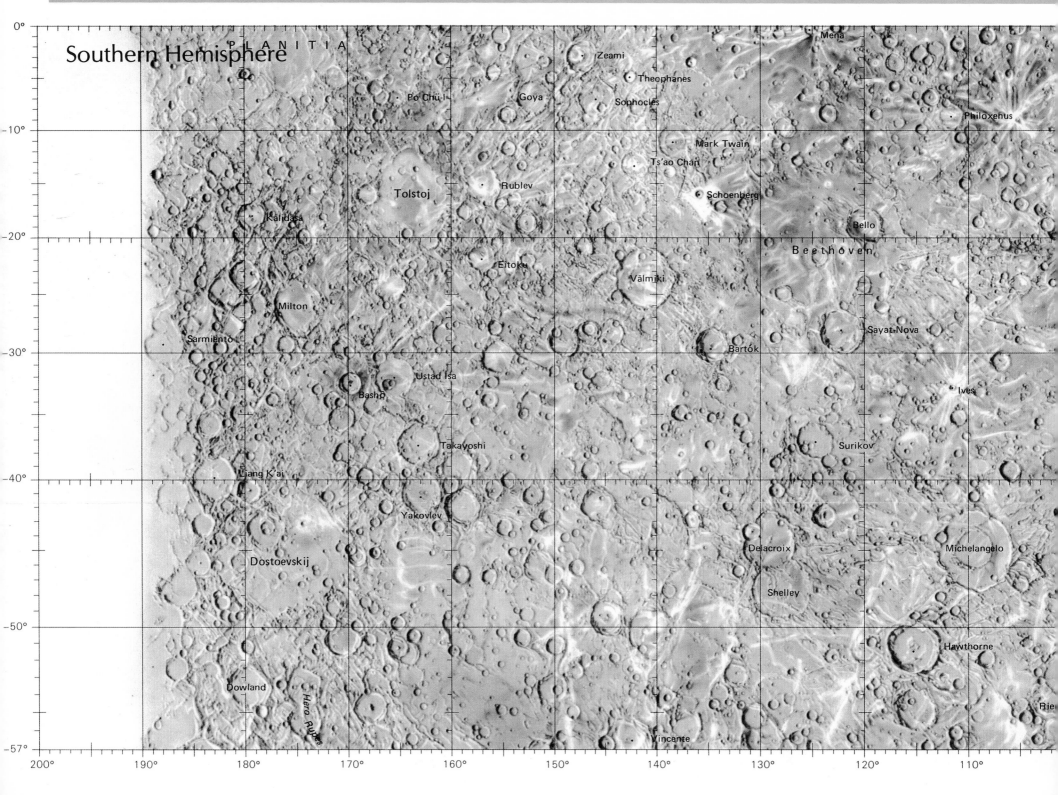

Southern Hemisphere

PLANITIA

Zeami
Theophanes
Po Chü-I
Goya
Sophocles
Mena
Philoxenus
Mark Twain
Ts'ao Chan
Rublev
Schoenberg
Tolstoj
Kalidasa
Bello
Beethoven
Eitoku
Vālmiki
Milton
Sayat-Nova
Sarmiento
Bartók
Ustad Isa
Basho
Ives
Takayoshi
Surikov
Liang K'ai
Yakovlev
Delacroix
Michelangelo
Dostoevskij
Shelley
Hawthorne
Dowland
Hero Rupes
Vincente
Rie

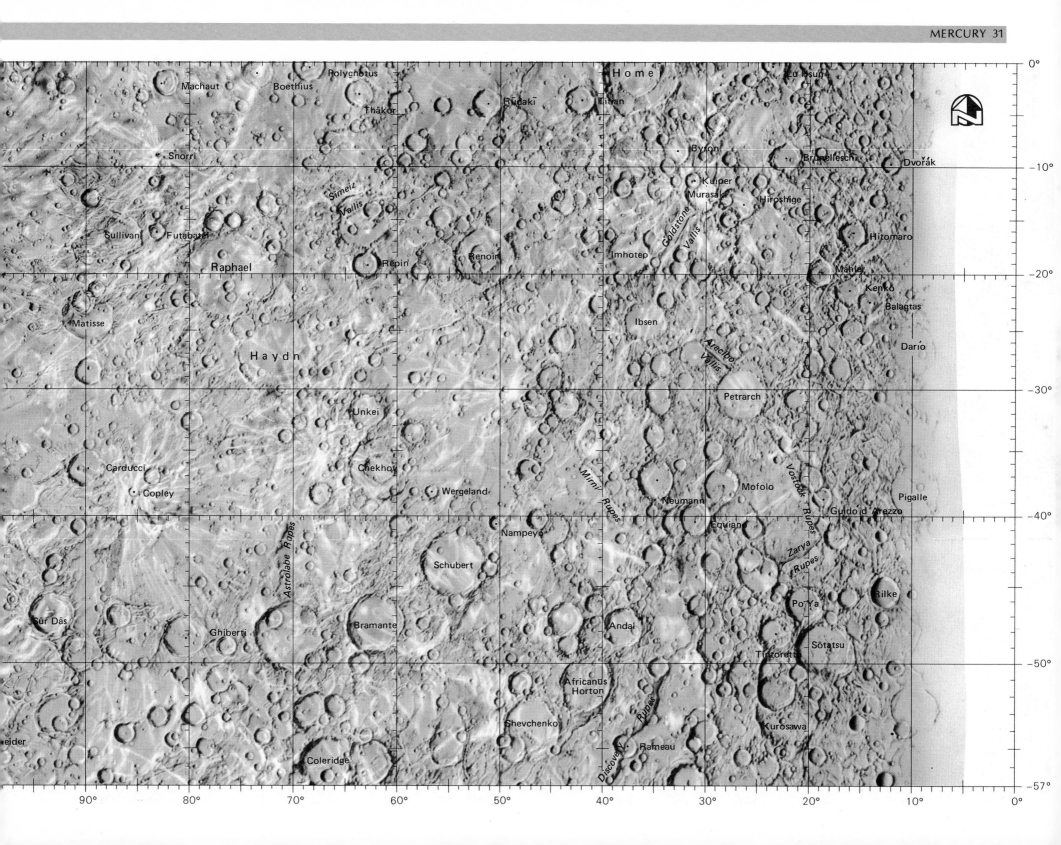

Machaut · Boethius · Polygnotus · Home · Lu Hsun · 0°

Thãkor · Rũdakĩ · Titian

Snorri · Byron · Brunelleschi · Dvořák · –10°

Simeiz · Kuiper

Vallis · Murasaki · Hiroshige

Sullivan · Futabatei · Goldstone · Hitomaro

Raphael · Imhotep · Vallis · Mahler · –20°

Rĕpin · Renoir · Kenkõ

Ibsen · Balagtas

Matisse · Dario

H a y d n · Arecibo

Vallis

Petrarch · –30°

Unkei

Chekhov · Vostock Rupes

Carducci · Mirni Rupes · Mofolo · Pigalle

Copley · Wergeland · Neumann · Guido d'Arezzo · –40°

Equiano

Astrolabe Rupes · Nampeyo · Zarya

Schubert · Rupes

Rilke

Po Ya

Sũr Dãs · Ghiberti · Bramante · Andal · Sõtatsu

Tintoretto

Africanus · –50°

Horton

Rupes

Shevchenko · Kurosawa

Discovery · Rameau

eider · Coleridge · Rupes · –57°

90° · 80° · 70° · 60° · 50° · 40° · 30° · 20° · 10° · 0°

VENUS

The second planet from the Sun, Venus is a near twin of the Earth in terms of size, with a diameter 95 percent that of our own planet. As viewed from Earth, Venus is the brightest celestial object in the sky except for the Sun and Moon. The Greek poet Homer even went so far as to call it the most beautiful star in the sky, while the Romans named it Venus after their goddess of beauty.

Like the Moon, Venus is seen to go through a series of phases as it orbits the Sun and is viewed from Earth. *Transits*—in which the planet passes directly between the Earth and the Sun —are characterized by an effect similar to an eclipse, although Venus appears as a mere tiny black dot creeping across the face of the Sun. Transits are rare, occurring in pairs eight years apart —and then not at all, for well over a century. The last pair of transits, for example, occurred in 1874-82, and the next will occur in 2004-12.

Early attempts at mapping the surface features of Venus were frustrated by the fact that the entire surface is covered by a thick cloud layer, a fact not known to early astronomers. Giovanni Cassini (1625-1712) produced the first 'map' in 1667, but as cloud patterns changed he could no longer find the features he had drawn. Johann Hieronymus Schroeter (1745-1816) was also fooled and reported having seen mountains on the surface. Schroeter, however, *was* the first to observe a very real phenomenon, that of the 'ashen light' seen in the Venusian atmosphere on the dark side of the planet. This faint light was at one time thought to be the city lights of Venusian civilization, but is now attributed to lightning which occurs during the planet's frequent electrical storms.

By the early twentieth century it had been determined that the Venusian surface was obscured by clouds, and various theories evolved regarding the actual nature of the surface beneath those clouds. The nineteenth century idea that the planet was covered by lush jungles was dismissed in favor of the two schools of thought that suggested either a vast desert or a vast ocean of water.

It had been established that the surface would be extremely hot because carbon dioxide in the thick atmosphere would prevent solar heat from escaping the surface, thus producing what is referred to as a 'greenhouse effect.'

The first successful expedition to the vicinity of Venus came in December 1962 when the American unmanned spacecraft Mariner 2 travelled to within 21,600 miles of the planet. The flight of Mariner 2 was a major milestone in unlocking the secrets of the mysterious planet. Among its achievements were confirmation that Venus has no detectable magnetic field, confirmation of the planet's exact rotational period—243 Earth days—and confirmation that it rotates from east to west rather than the opposite as previously supposed.

Mariner 2 also provided a more accurate reading of the planet's surface temperature, which at 900 degrees Fahrenheit is too hot for the existence of an ocean, because water could exist there only as steam. Water vapor is in fact present in the atmosphere and some astronomers have theorized that at an early stage in the evolution of Venus, oceans in fact *may have* existed on the surface.

In 1978, the United States undertook the Pioneer Venus project as a follow-on to several earlier Mariner probes. The project

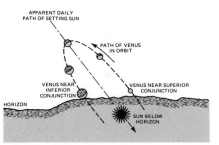

Facing page: This view of Venus prominently displays the planet's Y feature, which is created by Venus' atmospheric winds as they diverge at the planet's equator. The background has been added to illustrate how the planet would appear to the naked eye.

Venus' thick atmosphere is responsible for the reversed crescent (on the planet's left) shown in the *second frame above, at top.* This entire sequence depicts Venus in one of its rare transits of the face of the Sun. *Immediately above:* Venus' orbit around the Sun, as it is perceived by observers on Earth.

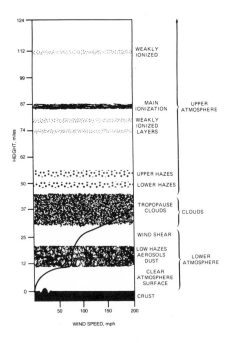

Above: A schematic diagram of the extremely dense Venusian atmosphere.

In March 1982, the Soviet Union's Venera 13 and 14 spacecraft transmitted the only color photos ever taken of the Venusian surface. The use of a super wide angle lens resulted in some distortion, but Venus' horizon can be seen in the upper corners of each photo *on the facing page—at top* is the left side of Venera 13's view, and *at bottom* is the total Venera 14 view.

The Venera cameras were destroyed by Venus' tremendous heat and pressure shortly after these photos were taken.

consisted of an orbiter spacecraft and a multiprobe spacecraft. The former undertook the detailed radar mapping of the Venusian surface that made possible the maps on these pages, and which gave us much of the information we now have about the planet's terrain. The multiprobe was actually five probes designed to return data about the Venusian atmosphere as they plunged toward the surface. One of the Pioneer Venus multiprobes continued to return data from the surface for just over an hour after impact.

The Soviet Union meanwhile prepared a series of spacecraft to conduct soft landings on the Venusian surface, which returned the only photographs ever taken of it. The Soviet Venera 9 and Venera 10 spacecraft each returned a single black and white

Data for Venus

Diameter: 7521 miles (12,104 km)
Distance from Sun: 67,580,000 miles (109,000,000 km) at
        aphelion
        66,588,000 miles (107,400,000 km) at
        perihelion
Mass: 2.213×10^{24} lb (4.8689×10^{24} kg)
Rotational period (Venusian day): 243.0 Earth days
        (retrograde)
Sidereal period (Venusian year): 224.7 Earth days
Eccentricity: 0.007
Inclination of rotational axis: 3°
Inclination to ecliptic plane (Earth = 0): 3.39°
Albedo (100% reflection of light = 0): .76
Mean surface temperature: 866.9° F
Maximum surface temperature: 900° F
Minimum surface temperature: 833° F
Highest point on surface: Maxwell Mountains
Largest surface feature: Aphrodite Terra
        6027.30 × 1988.39 miles
        (9700 × 3200 km)
Major atmospheric components: Carbon dioxide (96%)
Other atmospheric components: Nitrogen, Water vapor,
        Carbon monoxide,
        Hydrogen chloride,
        Hydrogen flouride,
        Sulphur dioxide
Atmospheric depth: 100 miles

image in 1975 and the Venera 13 and Venera 14 spacecraft returned color photos in March 1982.

While the Soviet Venera spacecraft provided the first photographs of specific points on the Venusian surface, the American Pioneer Venus Orbiter provided our first clear look at the overall global surface features of Venus. Using a radar altimeter, Pioneer Venus was able to obtain the data necessary to produce a topographical map of 90 percent of the planet's surface from 73 degrees north latitude to 63 degrees south latitude.

This data showed that the surface was generally smoother than the other three terrestrial planets with much less variation in altitude than is seen on earth. For instance, 60 percent of the Venusian surface is within 1600 feet of the planet's *mean radius* of 3752 miles. It has been suggested that this is due to the deeper lowlands having been filled with sand and other wind-blown material. Because there are no seas on Venus, the mean radius is used as a reference point in the same way that sea level is used on Earth.

Most of the surface of Venus is characterized as rolling uplands rising to an altitude of roughly 3000 feet while 20 percent of the surface is identified as lowlands and 10 percent as mountainous. The two largest upland regions or continental masses are Aphrodite Terra (roughly the size of Africa), near the equator in the southern hemisphere, and Ishtar Terra (roughly the size of Australia) in the northern hemisphere near the north pole. These two features constitute the Venusian 'continents' and are named respectively for the ancient Greek and ancient Babylonian goddesses of beauty.

The highest point on the mapped surface of Venus are the Maxwell Mountains (Maxwell Montes) in Ishtar Terra. High enough to have been identified by Earthbased radar prior to the Pioneer Venus project, the Maxwell Mountains, which may actually be a single mountain, rise to more than 35,000 feet above mean radius, or roughly 20 percent higher than Mount Everest rises above Earth's *sea level*. If viewed from the surface they would be an impressive sight, rising nearly 27,000 feet above Lakshmi Planvin, the surrounding plateau which is roughly the same elevation as the Tibetan plateau on earth.

Data obtained from Pioneer Venus indicates that the Maxwell Mountains may be the rim of an ancient volcano whose caldera had a diameter of roughly 60 miles. The lava flows, however, have long since been worn away by wind erosion and the slopes of the Maxwell Mountains are strewn with rocks and debris.

Another important upland region is Beta Regio with its great shield volcanos Rhea Mons and Theia Mons, which are larger

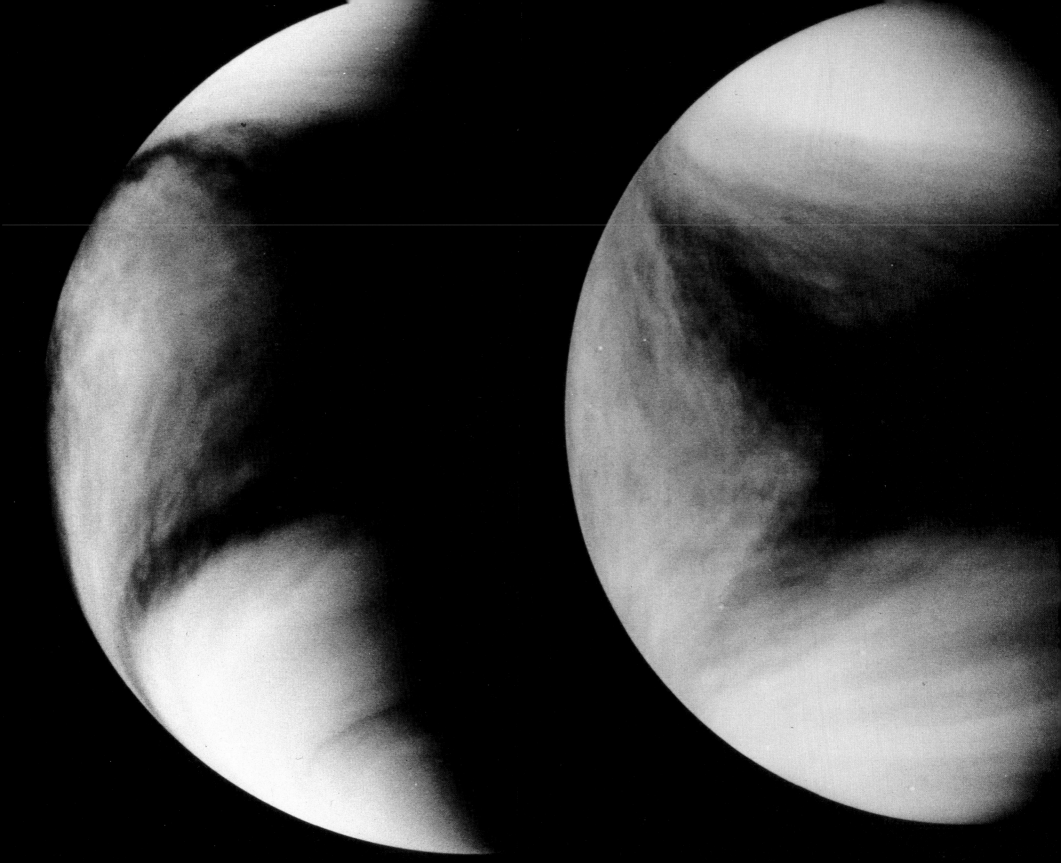

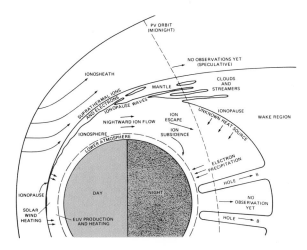

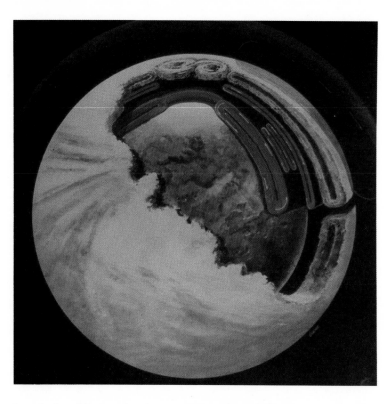

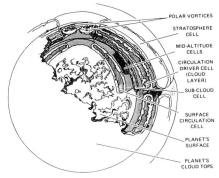

than the great shield volcanos of Hawaii on Earth. The mountainous Beta Regio is still in the process of formation and probably contains active volcanos. As such, it is the newest major surface feature on Venus.

The lowest point on the Venusian surface is actually a canyon, Diana Chasma, located within central Aphrodite Terra. At just 9500 feet below mean radius, Diana Chasma is much shallower than the corresponding lowest point on earth, the Marianas Trench. The largest and lowest lowland region on Venus is the Atalanta Plain (Atalanta Planitia) located northeast of Aphrodite Terra and due east of Ishtar Terra. It is roughly the same size as the Earth's North Atlantic Ocean although it is shallower by comparison.

The atmosphere of Venus has long been known to consist primarily of carbon dioxide, and the instruments of Pioneer Venus and Venera have pinpointed the proportion of carbon dioxide at 96 percent. Nitrogen constitutes more than three percent of the Venusian atmosphere and there are also traces of neon and several isotopes of argon.

There is some water vapor present in the Venusian cloud cover, where it has a density of 200 ppm—10 times the density of water vapor in the clear air near the surface. In the clouds the water vapor combines chemically with traces of sulphur dioxide to produce droplets of sulphuric acid, which give the Venusian cloud cover its distinctive yellowish color.

The Venusian cloud cover is complete and unbroken. The cloud layer is roughly 15 miles thick with its base about 30 miles above the surface of the planet, relatively higher than the thinner cloud cover on Earth. The air at the surface is probably quite clear and the air relatively still. The clouds, however, are pushed

by winds with speeds of up to 200 mph and circulate around the entire planet once every four Earth days, in contrast to the Venusian 'day' or rotation period of 243 Earth days.

Electric storms are common within the clouds and lightning has been detected by both American and Soviet spacecraft.

The most notable visible feature in the Venusian atmosphere, and one that misled so many would-be mapmakers in earlier days, is the Y Feature, whose tail sometimes stretches around the planet. The feature is actually the prevailing winds in the northern and southern hemisphere as they diverge at the equator. This pattern is constantly changing and sometimes it is seen as a reversed C. It always, however, retains an approximate north-south symmetry.

On the surface of Venus, atmospheric pressure is roughly 100 times that of the Earth. A yellowish glow like that of a smoggy sunset on Earth is all-pervasive. The daytime surface temperature of 900 degrees Fahrenheit is a global constant because the carbon dioxide/sulfuric acid atmosphere and cloud cover function as an insulating blanket, trapping the heat and producing convection currents that redistribute it across the entire surface area.

This page, extreme left: This diagram portrays the interaction of the solar wind with the Venusian atmosphere, as was determined from the experiments aboard the American Pioneer Venus spacecraft.

Facing page: The image of Venus *at left* was acquired by the Pioneer Venus spacecraft on 25 December 1978. The image *at right*, constructed from data acquired 16 days later, again shows the dark, horizontal Y feature. By studying Venus' weather, scientists hope to learn more about the weather on Earth.

Above left: A cutaway view of Venus, showing the complexity of its dense atmosphere and planetary surface; and *above*, a diagram of same, which names various constituent features.

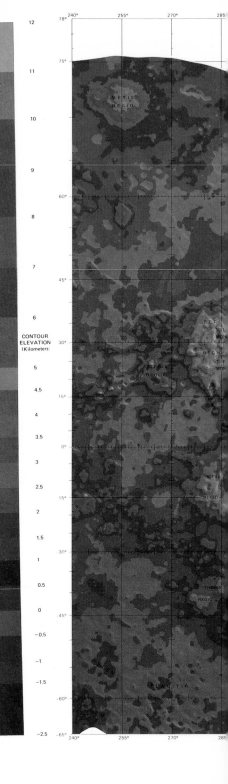

Above: Venus' northern 'continent,' Ishtar Terra, is shown at the top of this global view of Venus—which view had to be generated by the Pioneer spacecraft's radar altimeter, due to the planet's *visually* impenetrable cloud cover.

Right: This is how Ishtar Terra (shown here in approximately the same scale as on the map *on the facing page*) would appear if it were visible through Venus' opaque atmosphere, which it isn't—hence, the false color radar imaging of the photo and map.

CONTOUR ELEVATION (Kilometers)

Successful Earth Expeditions to Venus

Name	Country of Origin	Launch	Date of closest contact or landing	Distance of closest contact (miles)	Name	Country of Origin	Launch	Date of closest contact or landing	Distance of closest contact (miles)
Mariner 2	USA	27 Aug 62	14 Dec 62	21,595	Venera 11	USSR	9 Sep 78	25 Dec 78	Landed
Venera 4	USSR	12 Jun 67	18 Oct 67	Crushed in descent	Venera 12	USSR	14 Sep 78	25 Dec 78	Landed
Mariner 5	USA	14 Jun 67	19 Oct 67	2473	Venera 13	USSR	30 Oct 81	1 Mar 82	Landed
Venera 5	USSR	5 Jan 69	16 May 69	Crushed in descent	Venera 14	USSR	4 Nov 81	5 Mar 82	Landed
Venera 6	USSR	10 Jan 69	17 May 69	Crushed in descent	Vega 1	USSR	15 Dec 84	11 Jun 85	Lander reached surface, returned data for 21 min; balloon deployed at 54 km altitude, returned data for 46 hrs; 'bus' flew within 10,000 km of Halley's Comet nucleus 9 Mar 86
Venera 7	USSR	18 Aug 70	15 Dec 70	Reached surface; returned data for 23 min					
Venera 8	USSR	27 Mar 72	22 July 72	Reached surface; returned data for 50 min					
Mariner 10	USA	3 Nov 73	5 Feb 74	3571					
Venera 9	USSR	8 Jun 75	22 Oct 75	Reached surface; operated for 53 min					
Venera 10	USSR	14 Jun 75	25 Oct 75	Reached surface; survived 65 min	Vega 2	USSR	21 Dec 84	15 Jun 85	Lander analyzed soil samples; balloon returned data for 46 hrs; 'bus' flew within 7000 km of Halley's Comet nucleus 12 Mar 86
Pioneer-Venus 1	USA	20 May 78	4 Dec 78	Injected into orbit 4 Dec 78; orbital low point 93					
Pioneer-Venus 2	USA	8 Aug 78	9 Dec 78	Landed					

Radar Mercator Projection

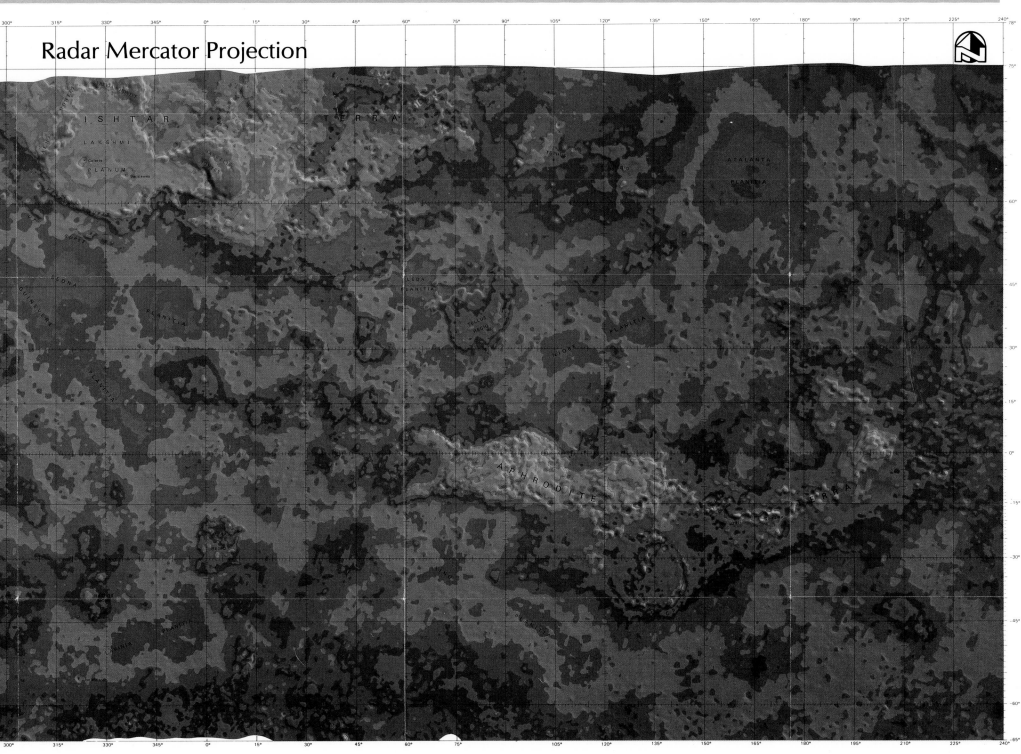

Northern Hemisphere

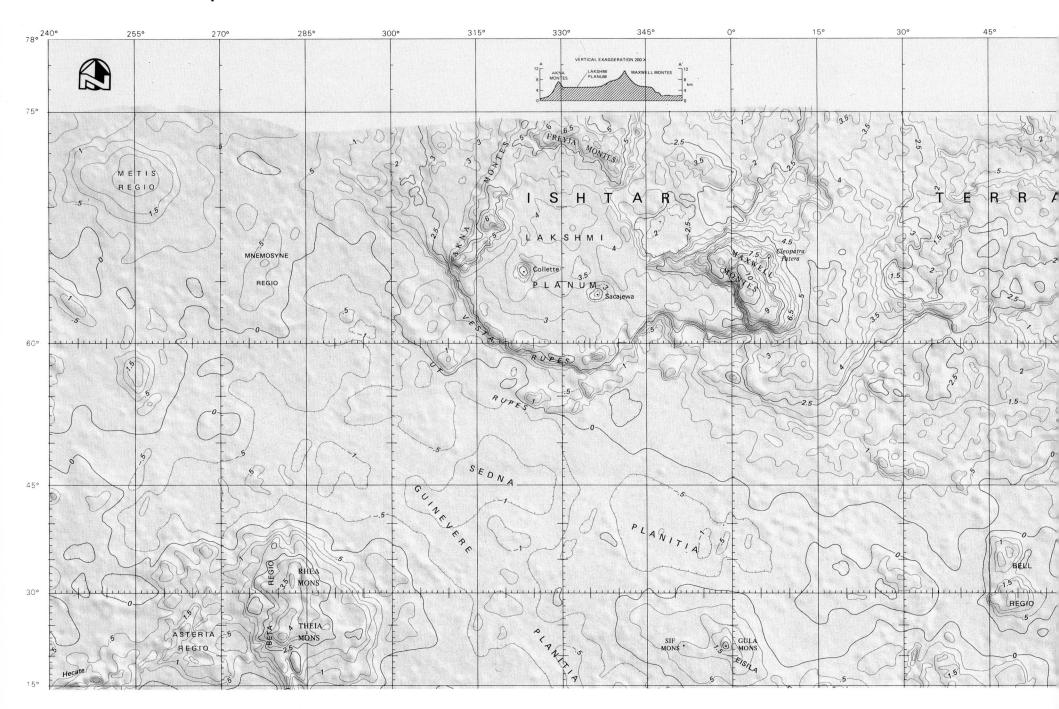

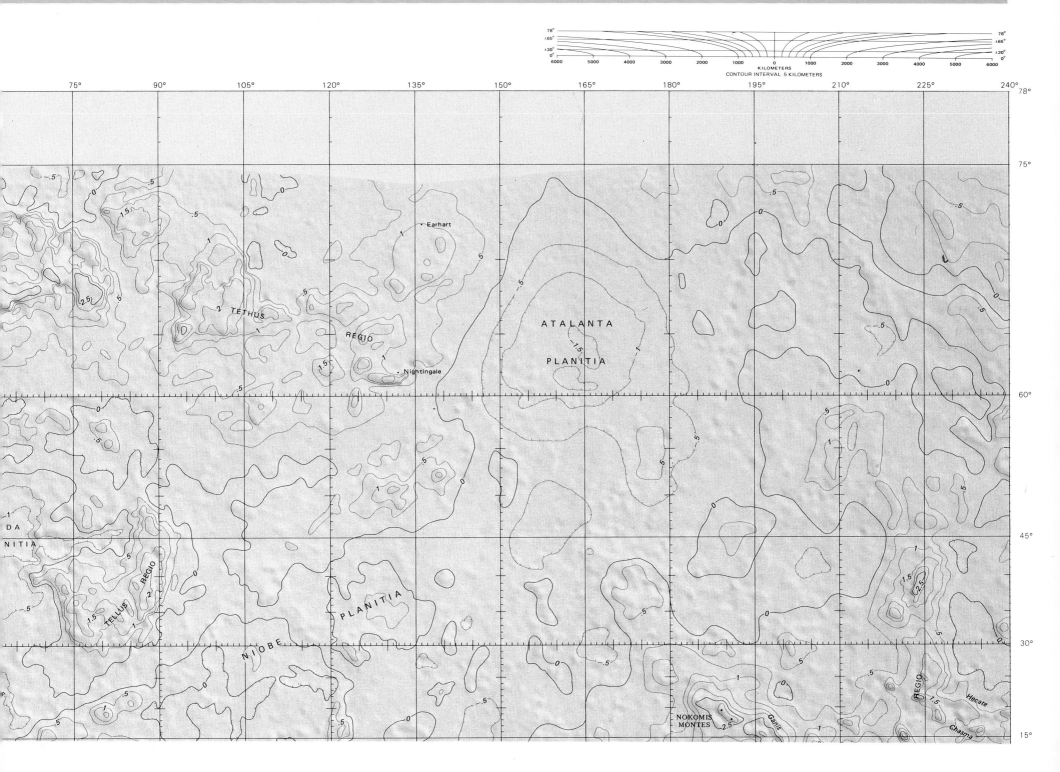

78°
±65°
±30°
0°

6000 5000 4000 3000 2000 1000 0 1000 2000 3000 4000 5000 6000
KILOMETERS
CONTOUR INTERVAL 5 KILOMETERS

75° 90° 105° 120° 135° 150° 165° 180° 195° 210° 225° 240°

78°

75°

• Earhart

TETHUS

REGIO

ATALANTA

• Nightingale

PLANITIA

60°

45°

D A

NITIA

TELLUS REGIO

PLANITIA

30°

NIOBE

REGIO

Hecate

NOKOMIS MONTES

Ganis

Chasma

15°

Southern Hemisphere

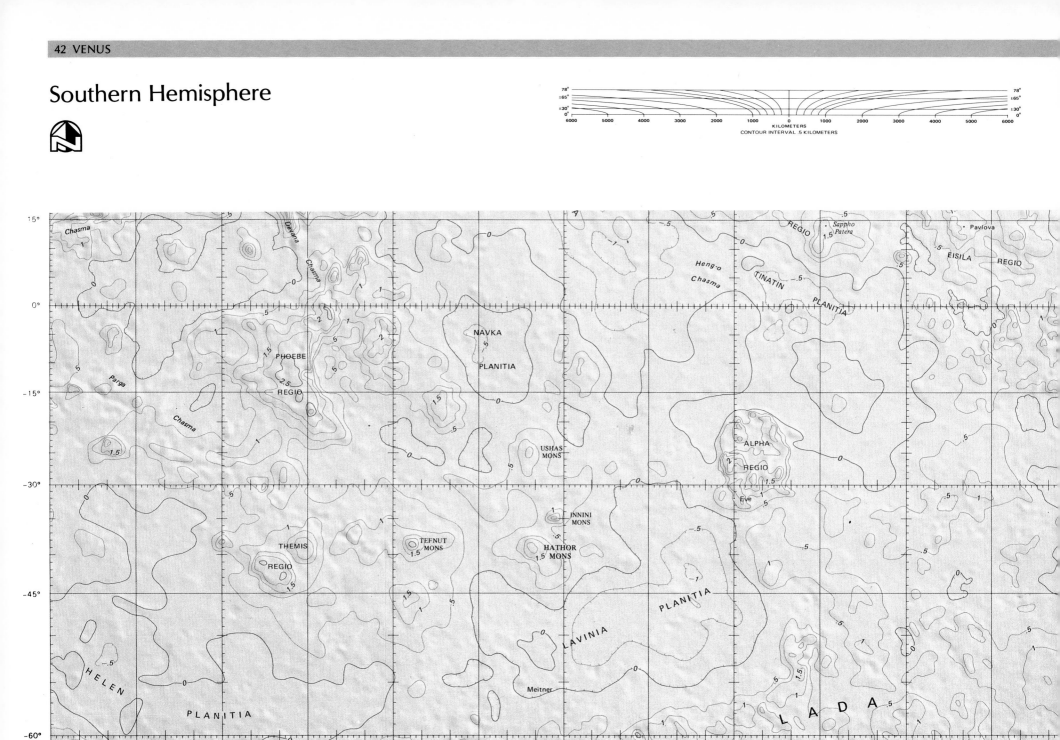

CONTOUR INTERVAL .5 KILOMETERS

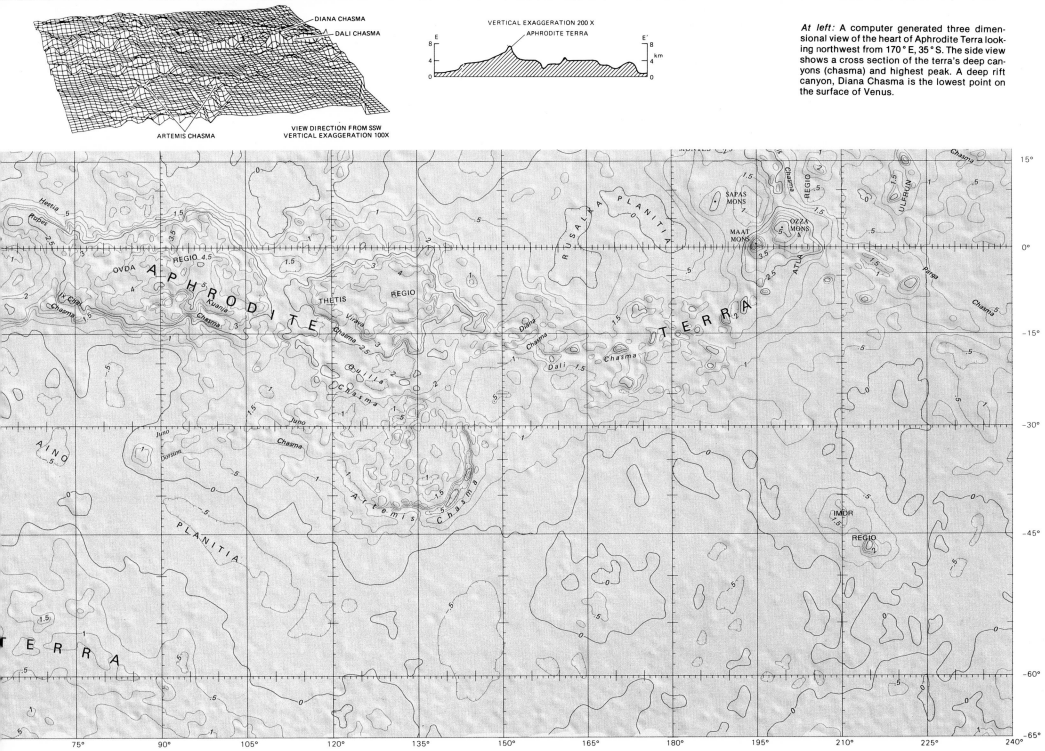

DIANA CHASMA
DALI CHASMA
ARTEMIS CHASMA

VIEW DIRECTION FROM SSW
VERTICAL EXAGGERATION 100X

VERTICAL EXAGGERATION 200 X
APHRODITE TERRA

At left: A computer generated three dimensional view of the heart of Aphrodite Terra looking northwest from 170° E, 35° S. The side view shows a cross section of the terra's deep canyons (chasma) and highest peak. A deep rift canyon, Diana Chasma is the lowest point on the surface of Venus.

EARTH

The third planet from the Sun, the Earth is the largest of the four terrestrial planets. The Earth is the only planet in the Solar System where life is known to have evolved, and because it is our home planet, we know more about its physical characteristics than we do of the other planets.

At the time the Solar System was formed 6.4 billion years ago, the Earth was probably solid throughout, but 500 million years later radioactive decay heated the Earth, and gradually metallic material melted and separated from non-metallic silicate material and sank toward the center of the Earth just as the silicates floated up. This molten metallic material, consisting mostly of iron with some nickel, survives today as the Earth's *core,* which is approximately 4200 miles in diameter. At temperatures of approximately 11,000 degrees Fahrenheit, the core is mostly molten, although a solid inner core perhaps 200 miles in diameter is thought to exist at the center of the Earth. The constant motion of the molten core gives the Earth its magnetic field.

Outside the core is the layer known as the *mantle.* Composed largely of both solid and molten silicate rock, the mantle is roughly 1800 miles thick. Covering the mantle is a thin *crust* which ranges from 25 to barely five miles thick. Occasionally, hot molten rock from the mantle forces its way through the crust in the form of lava during volcanic eruptions. So thin is the crust that it exists as a group of separate continental plates that literally float on the semi-liquid mantle, a phenomenon called *continental drift.* The edges of the continental plates are the faults and rift zones where volcanic activity and earthquakes are most common.

The Earth's major *continents* or land masses are, in order of size: Eurasia (21.3 million square miles), Africa (11.7 million square miles), North America (9.4 million square miles), South America (6.9 million square miles), Antarctica (six million square miles) and Australia (three million square miles). Prior to the Triassic period (two million years ago), the continents were combined in a single massive supercontinent called Pangaea. Over the next 160 million years a continent called Gondwana (Antarctica/Australia) drifted away from Pangaea on the semi-liquid mantle, and Eurasia/Africa gradually separated from North America/South America. Since the Cenozoic Period, 60 million years ago, Antarctica and Australia separated and the space between Eurasia/Africa and the Americas increased. It was by continental drift that the continents came into their present position on the Earth's surface, and it will be by this same process that the continents will continue to move apart.

The continents, however, are merely the uplands of the Earth's crust and together they comprise less than a third of the Earth's total surface area. Approximately 70 percent of the crust is covered by water in the form of four major oceans and a number of smaller bodies known as seas and lakes. The four major oceans are, in order of size, the Pacific (64 million square miles), the Atlantic (31.8 million square miles), the Indian (25.3 million square miles) and the Arctic (5.5 million square miles). The largest seas are the Mediterranean (1.2 million square miles) and Caribbean Sea/Gulf of Mexico (1.7 million square miles).

The Earth's oceans provide a convenient means of reckoning mean surface altitude (mean radius) which is referred to on Earth simply as *sea level.* The highest region on the continental mass of the Earth is the Tibetan plateau and the accompanying Himalaya Mountains which are located within the Eurasian continent. No fewer than the 47 tallest mountain peaks on Earth are located

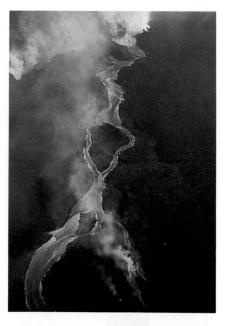

Facing page: Earth, as photographed from the returning US spacecraft Apollo 17. Visible here are the continents of Africa (top) and Antarctica (bottom). The background was added to show how the Earth would appear to the naked eye.

Above: A break in the Earth's crust lets lava flow from Hawaii's Mauna Loa volcano.

in this region. The tallest of these, Mount Everest, stands 5.5 miles above sea level. By comparison, the Maxwell Mountains on Venus rise to 6.6 miles above mean radius, while Mount Olympus on Mars towers 15.5 miles above mean radius.

The lowest point on the Earth is the Challenger Deep in the Marianas Trench of the Pacific Ocean, which is seven miles below sea level. Again, by comparison, the floor of the Diana Chasma on Venus is 1.8 miles below mean radius and the Hellas Basin is 1.9 miles below the mean radius of Mars.

Because of the effects of wind and water erosion, the meteor impact craters common to such other inner terrestrial bodies as Mercury, Mars and the Earth's own Moon are rare on the Earth. The effect of water alternately freezing and thawing also has a tendency to break up the rocks of the Earth's mountains. Consequently, the newly formed ranges on the Earth, such as the Rockies (which are between 70 and 300 million years old), tend to be higher than the more ancient ranges—such as the Appalachians, which are at least 400 million years old. The Earth's mountains were originally formed by the pressure of the continental plates moving against one another, and through volcanic action. Most of the ranges were formed by the former process, although the latter takes a relatively shorter time and is a good deal more spectacular. The sea mounts (mountains whose bases are on the ocean floor, but whose tops may be above sea level) of the Hawaiian Island chain are good examples of the latter, and the Island of Hawaii with its three active volcanos (Mauna Kea, Mauna Loa and Kiluea) is a good example of a sea mount that is still in the process of growing.

After the Jovian moon Io, the Earth is the second most volcanically active body in the Solar System with Venus being the only other body where volcanic activity is thought to be occurring at this time. The most volcanically active area on Earth is the Pacific basin, with volcanos active not only in Hawaii but in an arc around the north Pacific rim, stretching from Indonesia to Japan and Alaska and down the west coast of North America to the Cascades range—where the 1980 eruptions of Mount St Helens in Washington were very spectacular. Other volcanically active regions include the Italian peninsula and the North Atlantic (particularly Iceland).

Above the surface of the Earth is its *atmosphere,* which is comprised of several layers of gases roughly 120 miles thick and weighing 5700 trillion tons. Composed primarily of nitrogen (78 percent) and oxygen (21 percent), the Earth's atmosphere is divided into five layers. The *troposphere* is the densest and closest layer, covering the earth to a depth of seven miles. Next are

Data for Earth

Diameter: 7926 miles (12,756 km)

Distance from Sun: 94,240,000 miles (152,000,000 km) at aphelion

91,140,000 miles (147,000,000 km) at perihelion

Mass: 2.7155×10^{24} lb (5.9742×10^{24} kg)

Rotational period (Earth day): 23.93 Earth hours
23 hrs, 56 min

Sidereal period (Earth year): 365.2 Earth days

Eccentricity: 0.017

Inclination of rotational axis: 23.45°

Inclination to ecliptic plane (Earth = 0): 0°

Albedo (100% reflection of light = 1): .39

Mean surface temperature: 60.53° F

Maximum surface temperature: el Azizia, Libya 136.4° F
Death Valley, CA USA 134°F

Minimum surface temperature: Vostok, Antarctica −126.9° F

Highest point on surface: Mt Everest 29,028 ft

Largest surface feature: Pacific Ocean
64,186,000 sq mi
(166,883,600 sq km)

Major atmospheric components: Nitrogen (76.08%)
Oxygen (20.95%)
Argon (.934%)
Carbon dioxide (.031%)
Water (up to 1%)

Other atmospheric components: Neon, Helium, Methane, Krypton, Hydrogen, Nitrous oxide, Carbon monoxide, Xenon, Ozone

Atmospheric depth: 120 miles

the *stratosphere* or *ozonosphere* (7-30 miles), the *mesosphere* (30-50 miles) and the *ionosphere* (50 to 150 miles). Roughly 80 percent of the Earth's atmospheric molecules and most atmospheric pressure are concentrated in the troposphere. The atmosphere becomes much thinner above the mesosphere, so that 120 miles altitude is generally recognized as the edge of outer space. However, some remnants of Earthly atmosphere exist above 120 miles, so that the upper ionosphere and the region beyond is sub-divided into the *thermosphere* (60 to 400 miles) and *exosphere* (beyond 400 miles).

The Earth's atmosphere serves to shield the planet from much of the Sun's radiation. The visible spectrum penetrates all the atmospheric layers, but infrared and radio waves are partially blocked by the stratosphere. Ultraviolet radiation is almost entirely blocked by a layer of ozone in the stratosphere, and x-rays do not penetrate the mesosphere.

While the composition of the Earth's atmosphere is the same at all altitudes, clouds of water vapor are concentrated only in the troposphere. Here they are subdivided into low, middle and high clouds. Low clouds are typically the billowy *cumulus* as well as *stratus* and *nimbo stratus* types and exist up to an altitude of roughly 2.5 miles. Middle clouds are the thinner *altostratus* and patchy *altocumulus* types and exist between 2.5 and 4.5

Above: Fumes and steam rise from the river of hot lava pouring down Mauna Loa's side—evidence of past lava flows are conspicuous in a closer view of the volcano. See the detail view of Mauna Loa on page 45.

Facing page: Brooks Lake Creek Falls, in the Wyoming Rockies, evidences water and water erosion—features not found in the photos of Venus' surface on page 35, and of Mars' surface in the following chapter. The Rockies, formed when the continental plates collided, have been further shaped by eons of wind and water erosion. The water (foreground) reaches the mountains in cumulus clouds (background) and falls to the Earth in the form of rain, sleet or snow. As the mountains erode, their particles form soil in which plant life can grow, provided there is adequate sunlight and rain.

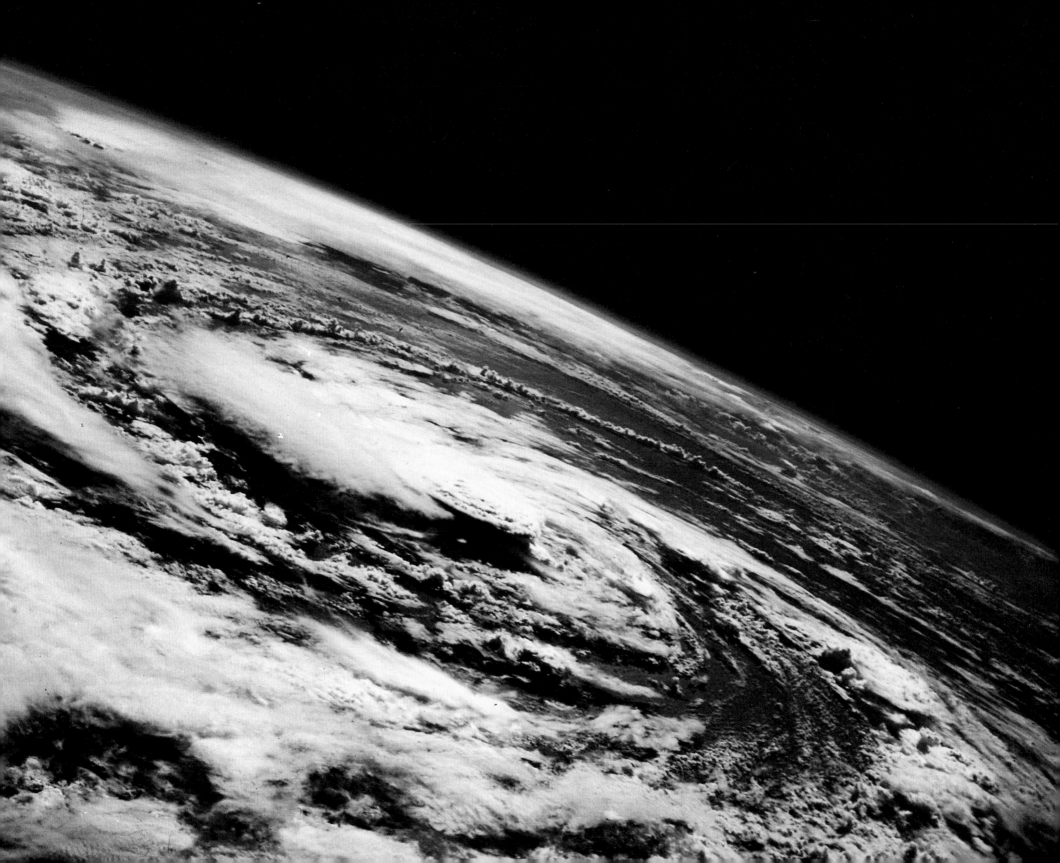

Above left: Shuttle Orbiter *Discovery* took this overhead photo of a heavy cumulus thunderhead formation—the blue areas visible in the cloud are lightning.

Facing page: This photograph of Hurricane Gladys, a cyclonic storm, was taken by the Apollo 7 spacecraft in October of 1968. The Earth's troposphere is visible as a light blue band on the horizon in this photo—and also on the horizon of the photo *above,* which gives evidence that the effects of water and wind erosion are very pronounced in the South Dakota Badlands. The rainbow, streaking upward toward the heavy cumulus clouds overhead, is the result of sunlight being refracted through water droplets which are here suspended as precipitation in the atmosphere—such a phenomenon is theoretically possible on any planet having an atmosphere which contains water vapor.

miles. High clouds include the thin *cirrostratus* and *cirronimbus* types as well as the distinctive wispy *cirrus.*

At any given moment half the Earth's surface is covered with clouds. This compares to 100 percent of the surface of Venus, less than 10 percent of the Martian surface and *none* of the surface of Mercury.

The Earth's cloud cover is constantly moving in global weather patterns. The Earth's water is always being recycled through the clouds. When clouds encounter cold air masses, water falls to Earth in the form of rain (liquid), snow (crystalline solid) or occasionally as hail (solid). Once on earth it flows into streams, and from there to rivers, and ultimately into the oceans and other large bodies of water, from which it evaporates back into the atmosphere to once again form a cloud.

During certain seasons, cyclonic storms (typhoons in the Pacific or hurricanes in the Atlantic) characterized by high winds can form in the Earth's cloud cover. By comparison, the cloud cover on Venus moves as a single unified mass with its constant Y Feature, while the cloud cover on Jupiter is characterized by numerous cyclonic storms. While such storms on Earth have a lifespan of a few days, however, those on Jupiter can last for dozens or even hundreds of years.

The Earth's atmosphere acts as a modulator of temperatures. Had the Earth been slightly warmer, it would have suffered the same sort of greenhouse effect that befell Venus and would today also have a carbon dioxide atmosphere.

The Earth's 23.4 degree inclination to its axis produces the *seasonal* effect, wherein the northern and southern hemispheres

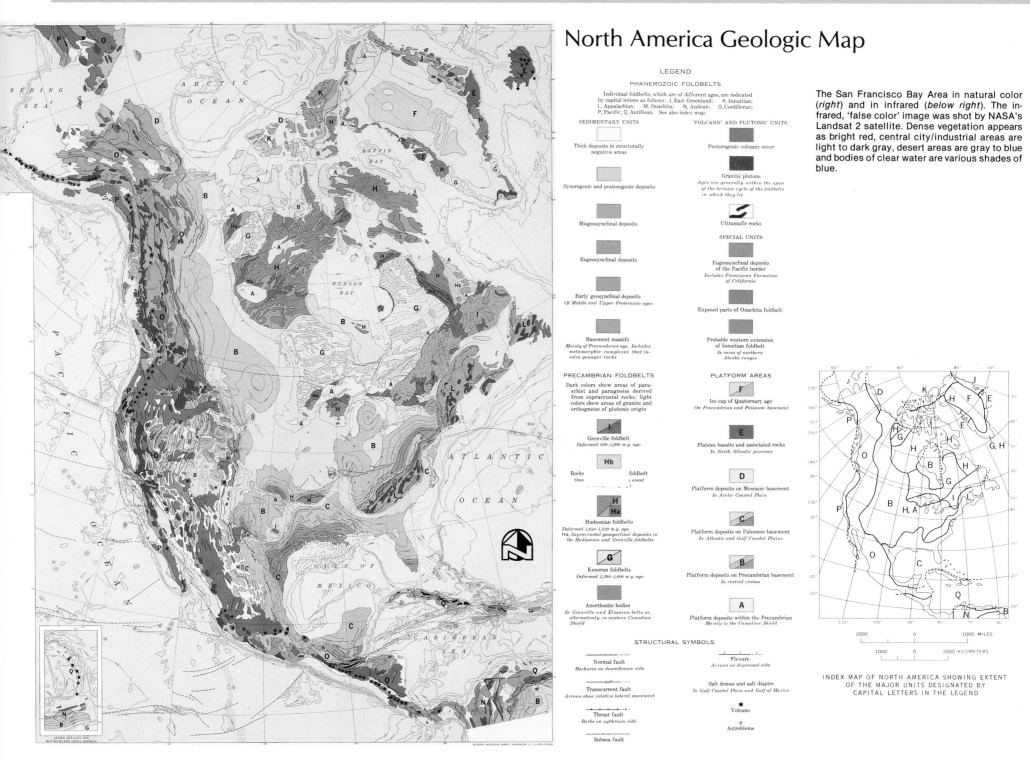

North America Geologic Map

LEGEND

PHANEROZOIC FOLDBELTS

Individual foldbelts, which are of different ages, are indicated by capital letters as follows: J, East Greenland; K, Innuitian; L, Appalachian; M, Ouachita; N, Andean; O, Cordilleran; P, Pacific; Q, Antillean. See also index map.

SEDIMENTARY UNITS

Thick deposits in structurally negative areas

Synorogenic and postorogenic deposits

Miogeosynclinal deposits

Eugeosynclinal deposits

Early geosynclinal deposits
Of Middle and Upper Proterozoic ages

Basement massifs
Mainly of Precambrian age. Includes metamorphic complexes that involve younger rocks

VOLCANIC AND PLUTONIC UNITS

Postorogenic volcanic cover

Granitic plutons
Ages are generally within the span of the tectonic cycle of the foldbelts in which they lie

Ultramafic rocks

SPECIAL UNITS

Eugeosynclinal deposits of the Pacific border
Includes Franciscan Formation of California

Exposed parts of Ouachita foldbelt

Probable western extension of Innuitian foldbelt
In cores of northern Alaska ranges

PRECAMBRIAN FOLDBELTS

Dark colors show areas of paraschist and paragneiss derived from supracrustal rocks; light colors show areas of granite and orthogneiss of plutonic origin

Grenville foldbelt
Deformed 880–1,000 m.y. ago

Hb Rocks   foldbelt
Over

H **Ha** Hudsonian foldbelts
Deformed 1,640–1,820 m.y. ago
Ha, Supracrustal geosynclinal deposits in the Hudsonian and Grenville foldbelts

G Kenoran foldbelts
Deformed 2,390–2,600 m.y. ago

Anorthosite bodies
In Grenville and Elsonian belts or, alternatively, in eastern Canadian Shield

PLATFORM AREAS

F Ice cap of Quaternary age
On Precambrian and Paleozoic basement

E Plateau basalts and associated rocks
In North Atlantic province

D Platform deposits on Mesozoic basement
In Arctic Coastal Plain

C Platform deposits on Paleozoic basement
In Atlantic and Gulf Coastal Plains

B Platform deposits on Precambrian basement
In central craton

A Platform deposits within the Precambrian
Mainly in the Canadian Shield

STRUCTURAL SYMBOLS

Normal fault
Hachures on downthrown side

Transcurrent fault
Arrows show relative lateral movement

Thrust fault
Barbs on upthrown side

Subsea fault

Flexure
Arrows on depressed side

Salt domes and salt diapirs
In Gulf Coastal Plain and Gulf of Mexico

★ Volcano

⊕ Astrobleme

The San Francisco Bay Area in natural color (*right*) and in infrared (*below right*). The infrared, 'false color' image was shot by NASA's Landsat 2 satellite. Dense vegetation appears as bright red, central city/industrial areas are light to dark gray, desert areas are gray to blue and bodies of clear water are various shades of blue.

INDEX MAP OF NORTH AMERICA SHOWING EXTENT OF THE MAJOR UNITS DESIGNATED BY CAPITAL LETTERS IN THE LEGEND

are alternately closest to the Sun. Only twice each Earth year, on the *vernal* and *autumnal equinoxes*, will the Sun shine directly on the Earth's equator. In the northern hemisphere, for example, the inclination toward the Sun increases from the *vernal equinox* (21 March) until the *summer solstice* (21 June). At the summer solstice, the northern hemisphere is oriented so that it receives the most sunlight of the year, while the southern hemisphere receives the least. From the summer solstice until the *autumnal equinox* (23 September), the amount of sunlight gradually decreases in the northern hemisphere, and increases in the southern as the Sun is perceived to cross the equator. On the autumnal equinox the Sun 'crosses' the equator and the trend continues until the *winter solstice* (21 December), when the most sunlight reaches the southern hemisphere and the least sunlight reaches the northern hemisphere. After the winter solstice, the Sun's warmth once again moves north toward the equator and the northern hemisphere, and the seasonal cycle goes around again and again. The Sun is directly overhead only at the equinoxes. This regular annual pattern determines the *climate* of all the regions of the Earth.

Changes in the *axial inclination,* which varies by as much as 2.5 degrees in 100,000 years, can produce such dramatic changes in climate as the Ice Age, which covered 25 percent of the Earth's land area with ice until 15,000 years ago.

Throughout the Earth's annual seasonal cycle, its equator receives more sunlight than any other part of the planet, while the poles receive the least. For this reason, the Earth's poles have permanent ice caps which recede slightly with their respective warmer seasons, but which are always present. In the south polar region there is the ice-covered continent Antarctica, while in the north polar region there is the ice-covered Arctic Ocean. The Antarctic ice cap contains an average 6.3 million cubic miles of frozen water, while the Arctic ice cap and associated North American and Eurasian glaciers (including Greenland) contain 680,000 cubic miles of water.

The inclination of the Earth's orbit to the Earth's axis and the resulting seasonal effect produce annual changes in temperature and pressure. These in turn control the Earth's global weather pattern and the movement of the clouds. The mountains on the Earth's continents can in turn affect weather by triggering precipitation on their windward side.

The complex interrelationship of axial inclination, chemical composition and geographical composition that is unique to the Earth probably played an important role in the development of the phenomena of life.

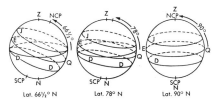

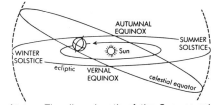

At top: The diurnal path of the Sun, as evidenced by what might be called its 'ground tracks' on Earth—from its farthest north in June (J), to its farthest south in December (D), in relation to the horizon (broken lines), at latitudes 66½°, 78° and 90° north. Z and N equal the observer's zenith and nadir, NCP and SCP the north and south celestial poles and EQ the celestial equator. The arrow indicates polar attitude.

Above: A chart of the Earth's equinoctial relationship with the Sun. 'Equinox' signifies equality of day and night; solstice indicates 'standing'—as when the Sun has no apparent northward or southward motion—and further implies an inequality of day and night: in summer, more day than night; in winter the reverse.

Western Arctic Region

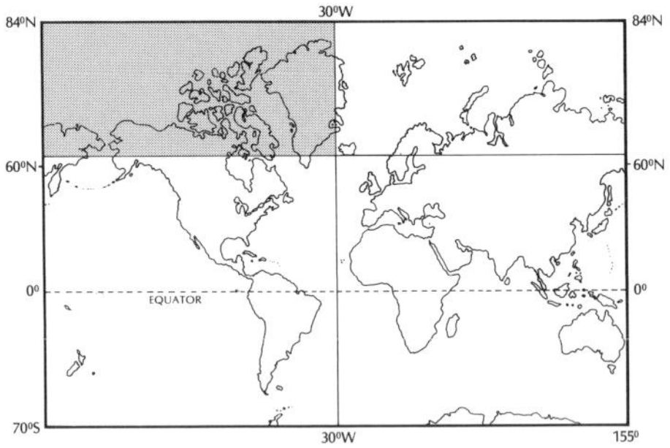

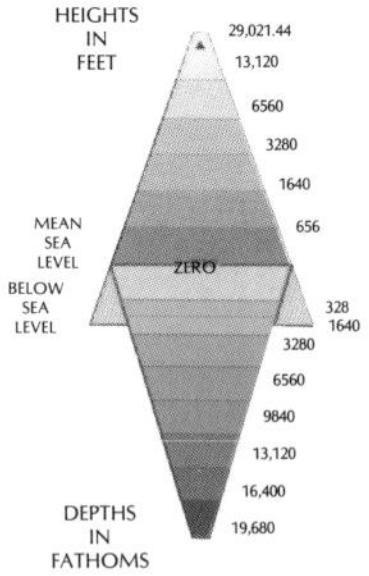

84°N   30°W

60°N   60°N

0°   EQUATOR   0°

70°S   30°W   155°

HEIGHTS IN FEET

	29,021.44
	13,120
	6560
	3280
	1640
	656
MEAN SEA LEVEL	
BELOW SEA LEVEL	ZERO
	328
	1640
	3280
	6560
	9840
	13,120
	16,400
DEPTHS IN FATHOMS	19,680

Earth's western Arctic region is frequently under cloud cover, but in the photograph *below,* Greenland's ice cap can be seen through breaks in the clouds. This photo was taken in April of 1972, and was shot from aboard Apollo 16 during its return from the fifth successful US manned lunar landing. Among the Solar System's planets, only the Earth and Mars have visible polar ice caps.

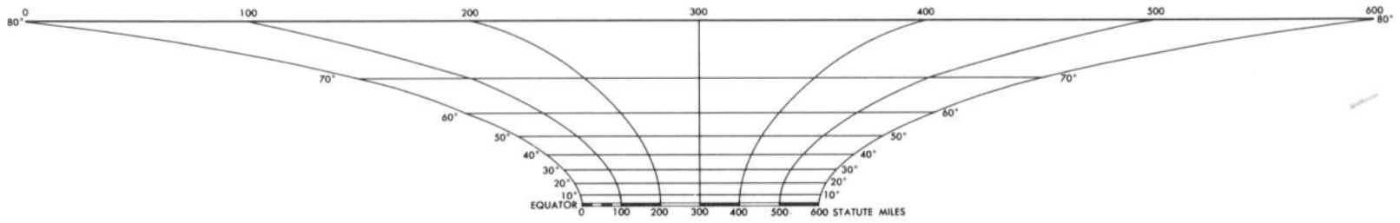

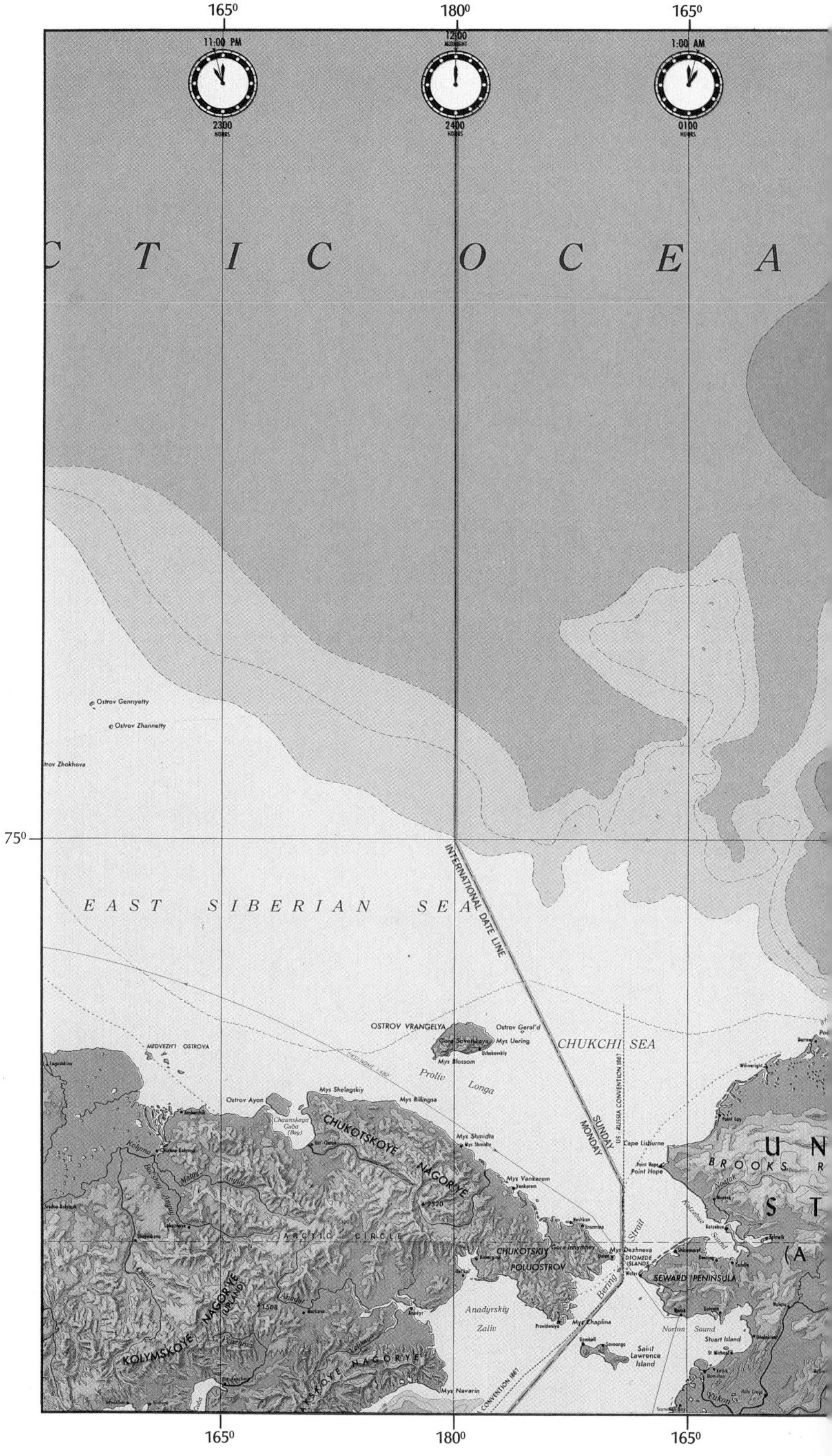

165°   180°   165°

11:00 PM   12:00 MIDNIGHT   1:00 AM
2300 HOURS   2400 HOURS   0100 HOURS

A R C T I C   O C E A

75°

E A S T   S I B E R I A N   S E A

INTERNATIONAL DATE LINE

Ostrov Gennyaty
Ostrov Zhannetty
Ostrov Zhokhova

MEDVEZHI OSTROVA

OSTROV VRANGELYA

Mys Uering
Ostrov Geral'd

CHUKCHI SEA

Proliv   Longa

Ostrov Ayon   Mys Shelagskiy   Mys Bilinga

Chaunskaya Guba (Bay)

CHUKOTSKOYE   NAGORYE

Mys Shmidta

Cape Lisburne

Point Hope

B R O O K S

U N
S T
(A

SEWARD PENINSULA

KOLYMSKOYE   NAGORYE

CHUKOTSKIY
POLUOSTROV

DIOMEDE ISLANDS

Bering   Strait

Anadyrskiy
Zaliv

Saint Lawrence Island

Norton
Sound

ARCTIC CIRCLE

165°   180°   165°

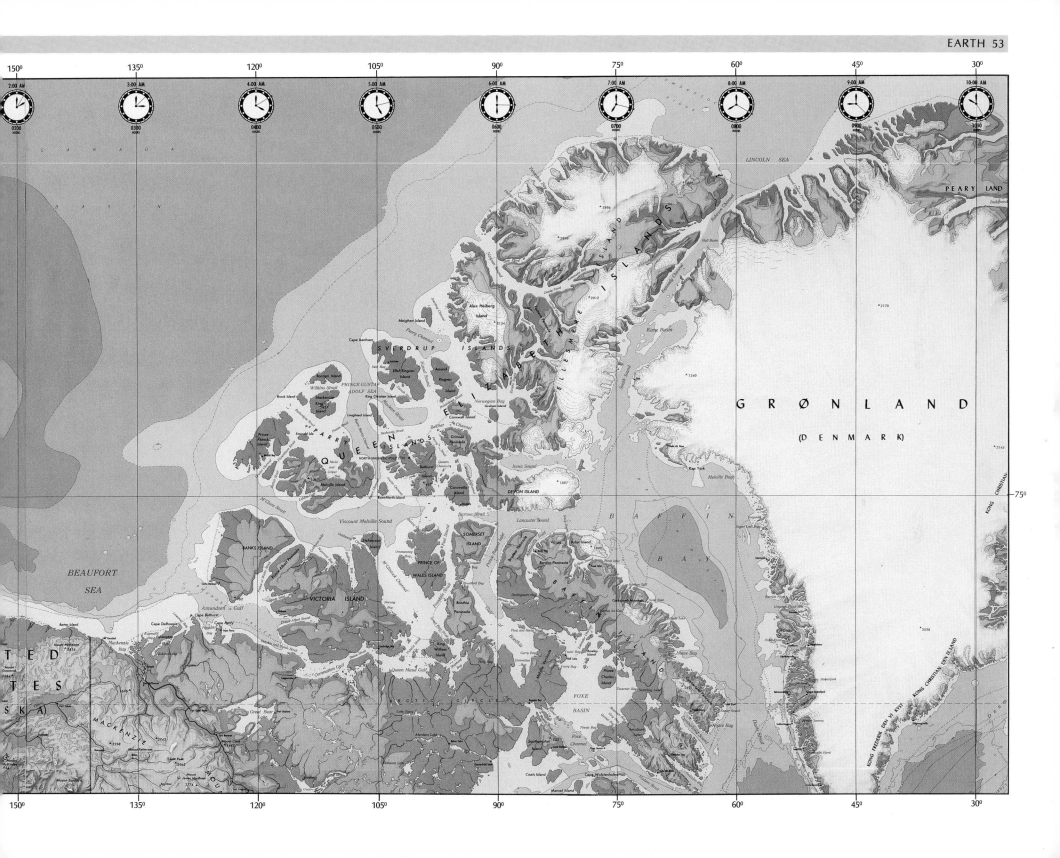

North Pacific Ocean and Northern Western Hemisphere

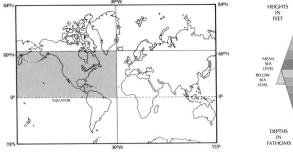

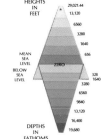

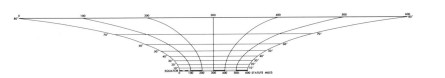

Below: Landsat 4's Thematic Mapper instrumentation provided this high relief view of Death Valley, here surrounded by (clockwise from left), the Panamint Range; the Grapevine, Funeral and Black Mountains; the Spring Mountains and Las Vegas Valley; Searles Lake; and the Granite and Avawatz Mountains in the northern Mojave Desert. The valleys and basins shown in this photograph contain fine, light toned clays and associated salt deposits. Several major fault zones pass through this area.

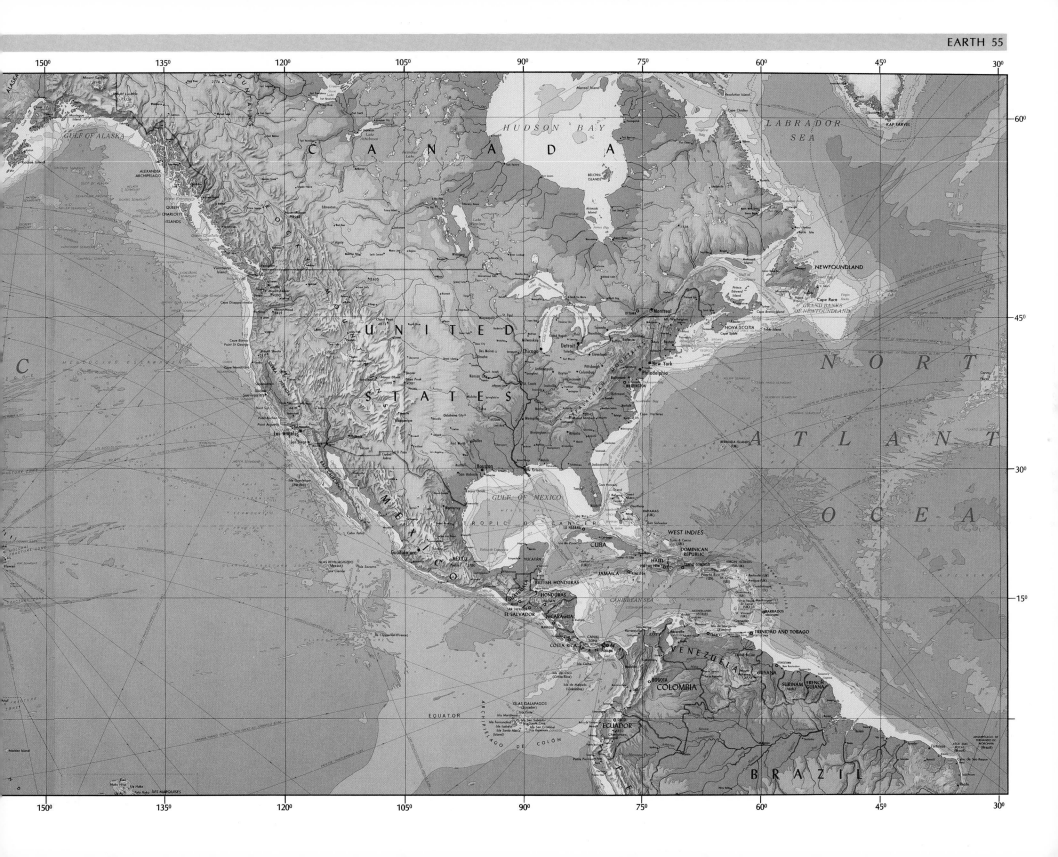

South Pacific Ocean and Southern Western Hemisphere

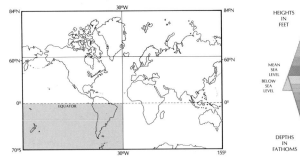

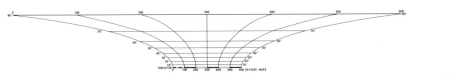

HEIGHTS
IN
FEET

29,021.44
13,120
6560
3280
1640
656

MEAN
SEA
LEVEL

ZERO

BELOW
SEA
LEVEL

328
1640
3280
6560
9840
13,120
16,400
19,680

DEPTHS
IN
FATHOMS

Below: This view of a high intensity cumulon-imbus formation over the Amazon Basin in Brazil was taken by Apollo 9 in 1969. The Amazon Basin receives some of the Earth's heaviest precipitation, and is an important weather formation area due to its vast, low-reflective tree cover.

Inset, opposite: A Skylab 3 photo of a snow-capped volcano in New Zealand.

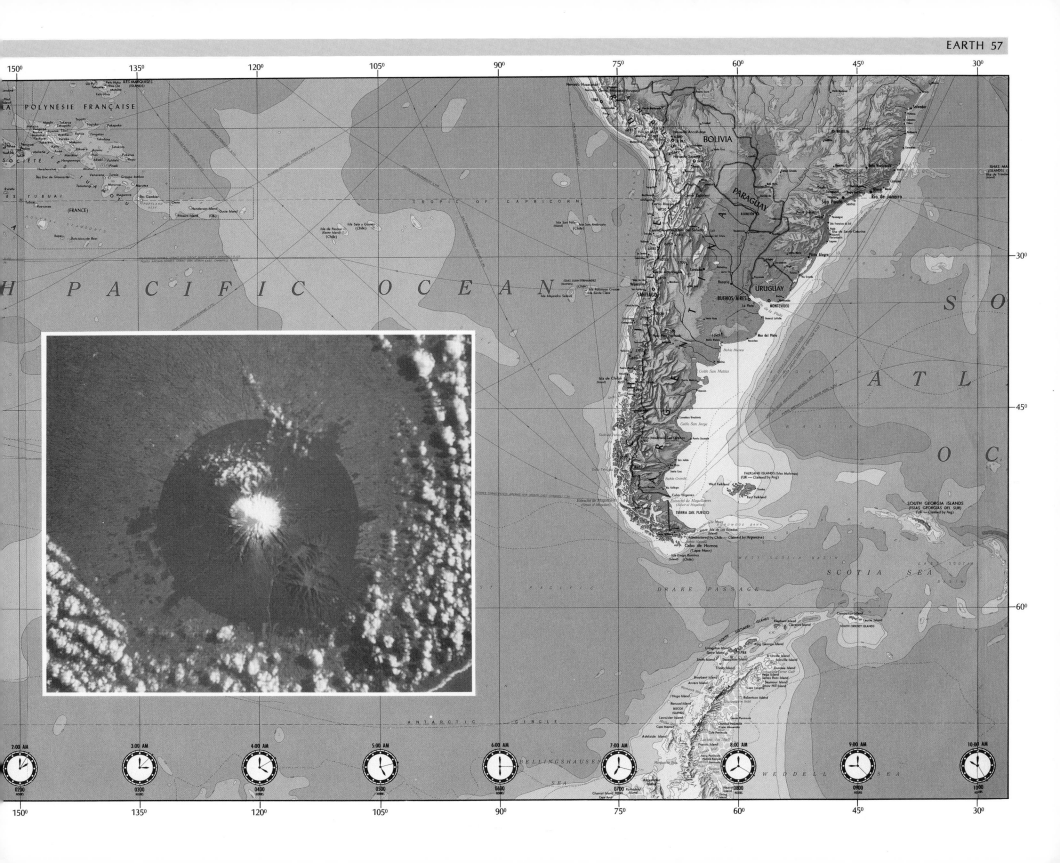

Eastern Arctic Region

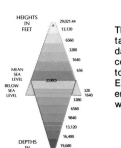

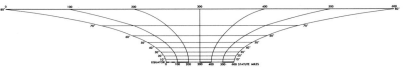

HEIGHTS
IN
FEET

29,021.44
13,120
6560
3280
1640
656

MEAN
SEA
LEVEL
ZERO

BELOW
SEA
LEVEL

328
1640
3280
6560
9640
13,120

DEPTHS
IN
FATHOMS

16,400
19,680

This photo (*below*) of atmospheric vortices was taken by Gemini 10 during that craft's three-day voyage in July of 1966. Evident here is the counterclockwise spiralling motion imparted to the northern hemispheric winds by the Earth's own rotational movement. In the southern hemisphere, analogous vortice motion would be clockwise.

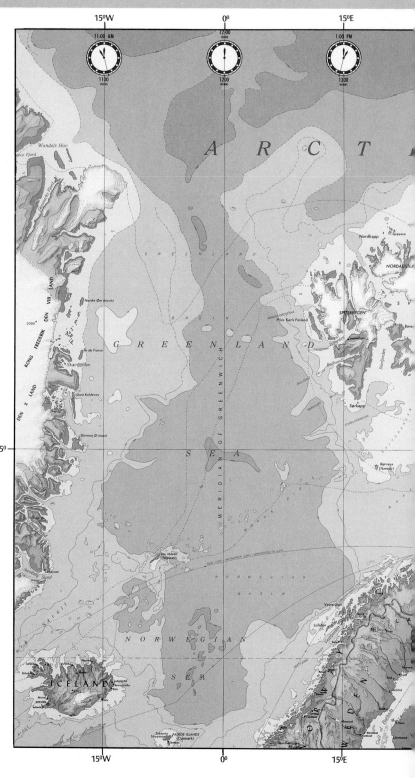

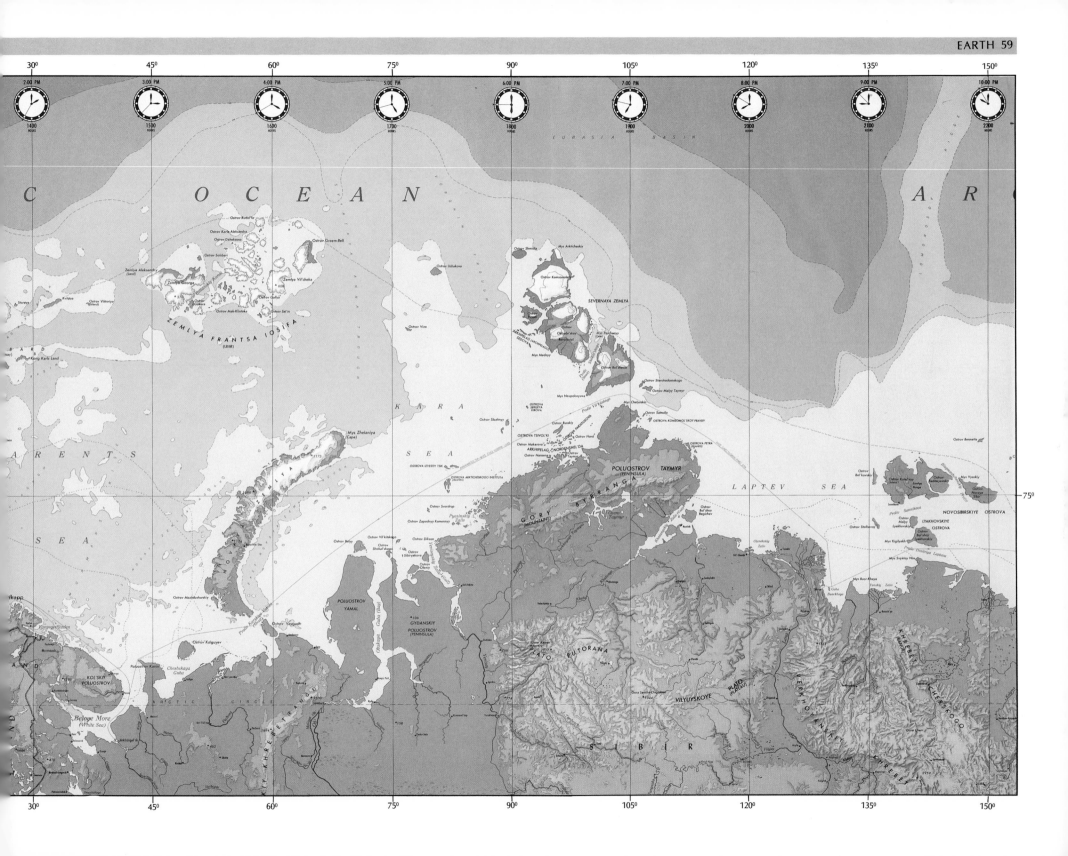

Eurasia and Northern Africa

HEIGHTS
IN
FEET

MEAN
SEA
LEVEL

BELOW
SEA
LEVEL

DEPTHS
IN
FATHOMS

Below: NASA's Landsat 1 Earth resources satellite took this photo of the Himalaya Mountains in the eastern third of the mountain kingdom of Nepal. Formed when the Indian subcontinent slammed into Asia, this mountain range includes Mount Everest—at 29,028 feet, the highest point on the face of the Earth.

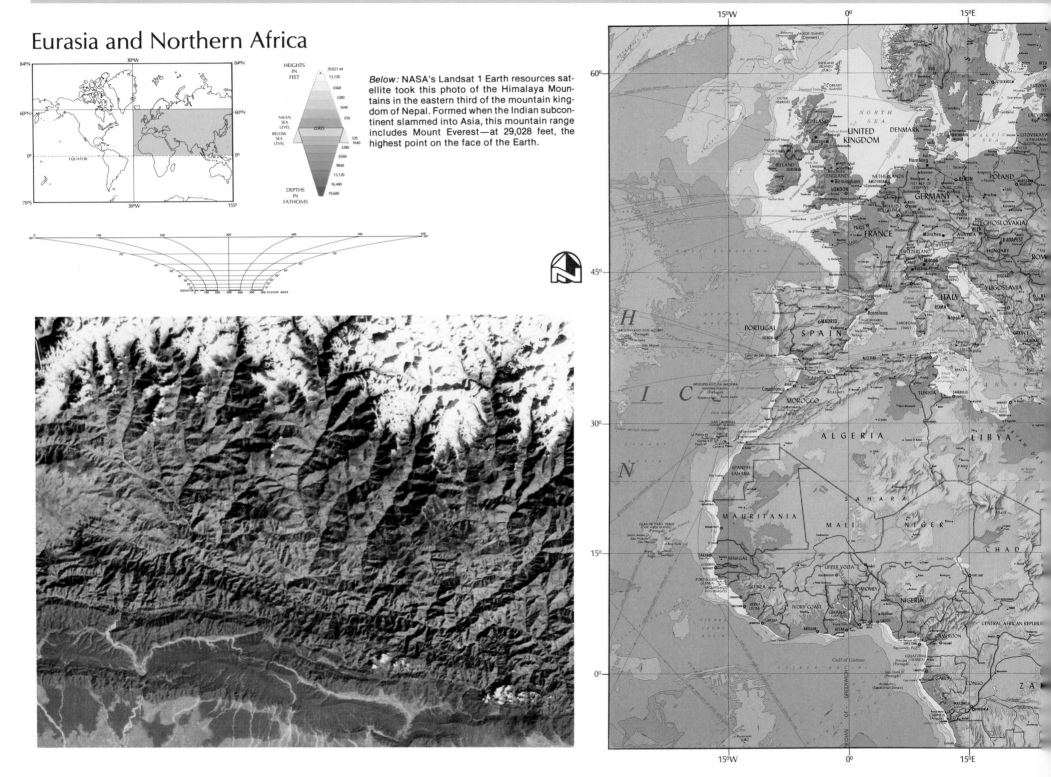

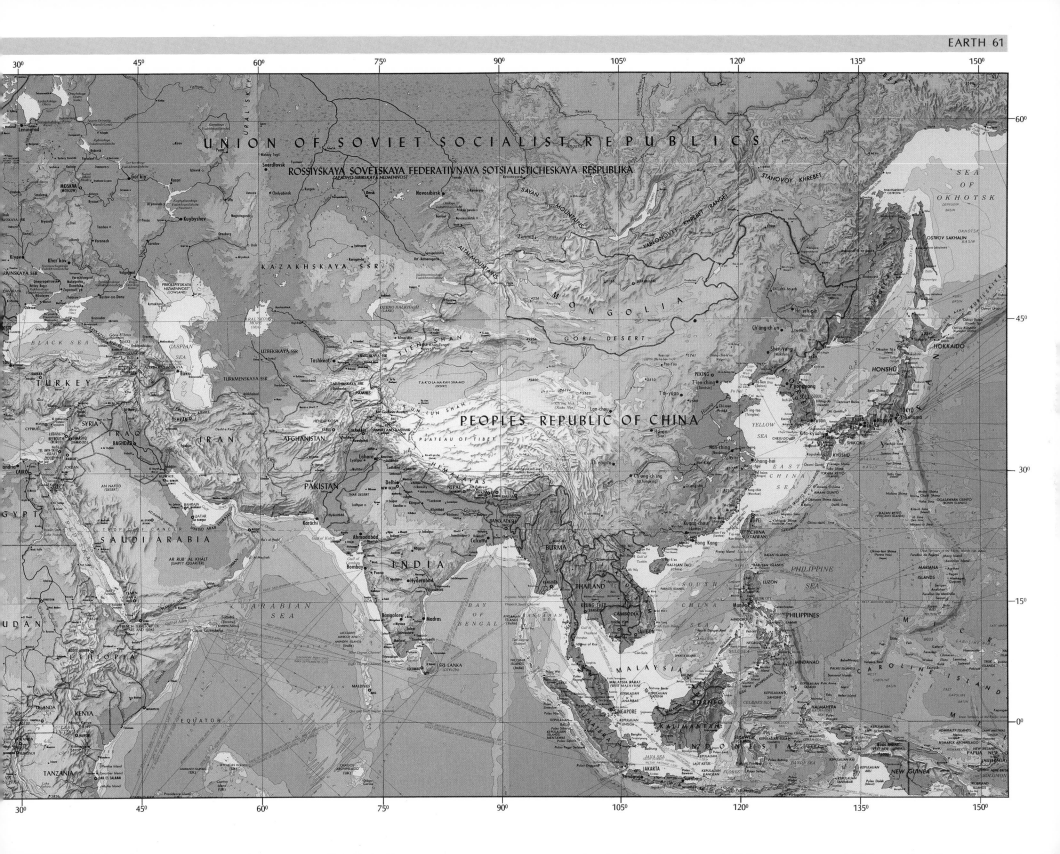

Indian Ocean, Southern Africa and Australia

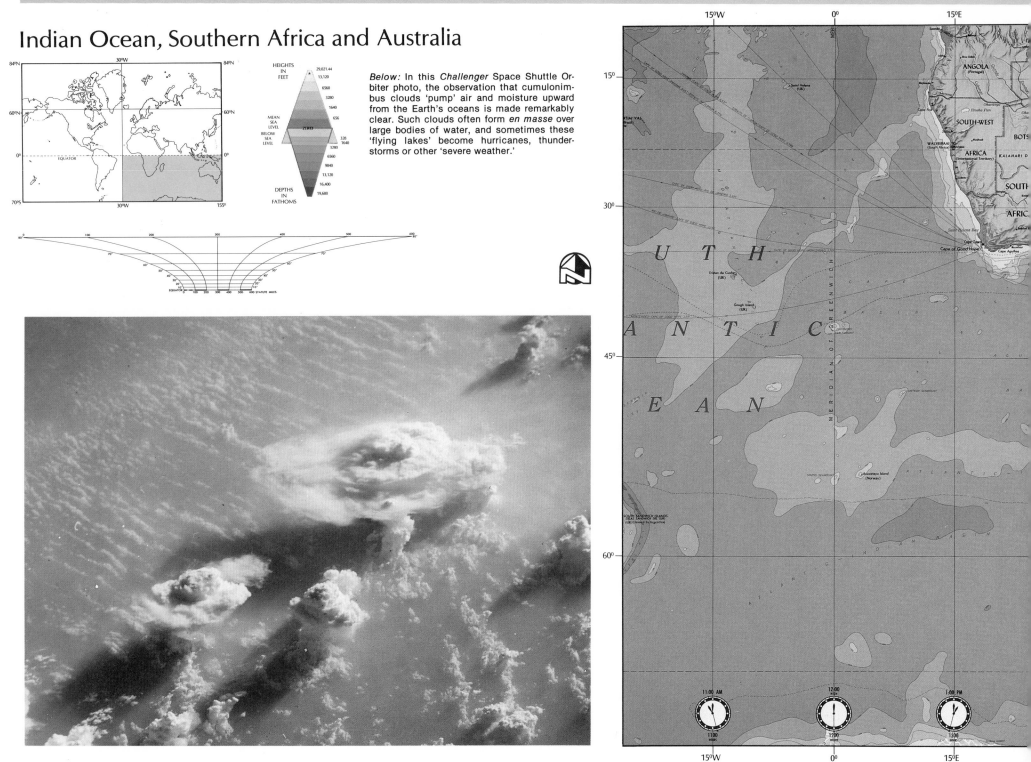

Below: In this *Challenger* Space Shuttle Orbiter photo, the observation that cumulonimbus clouds 'pump' air and moisture upward from the Earth's oceans is made remarkably clear. Such clouds often form *en masse* over large bodies of water, and sometimes these 'flying lakes' become hurricanes, thunderstorms or other 'severe weather.'

HEIGHTS IN FEET

29,021.44
13,120
6560
3280
1640
656

MEAN SEA LEVEL

ZERO

BELOW SEA LEVEL

328
1640
3280
6560
9840
13,120
16,400

DEPTHS IN FATHOMS

19,680

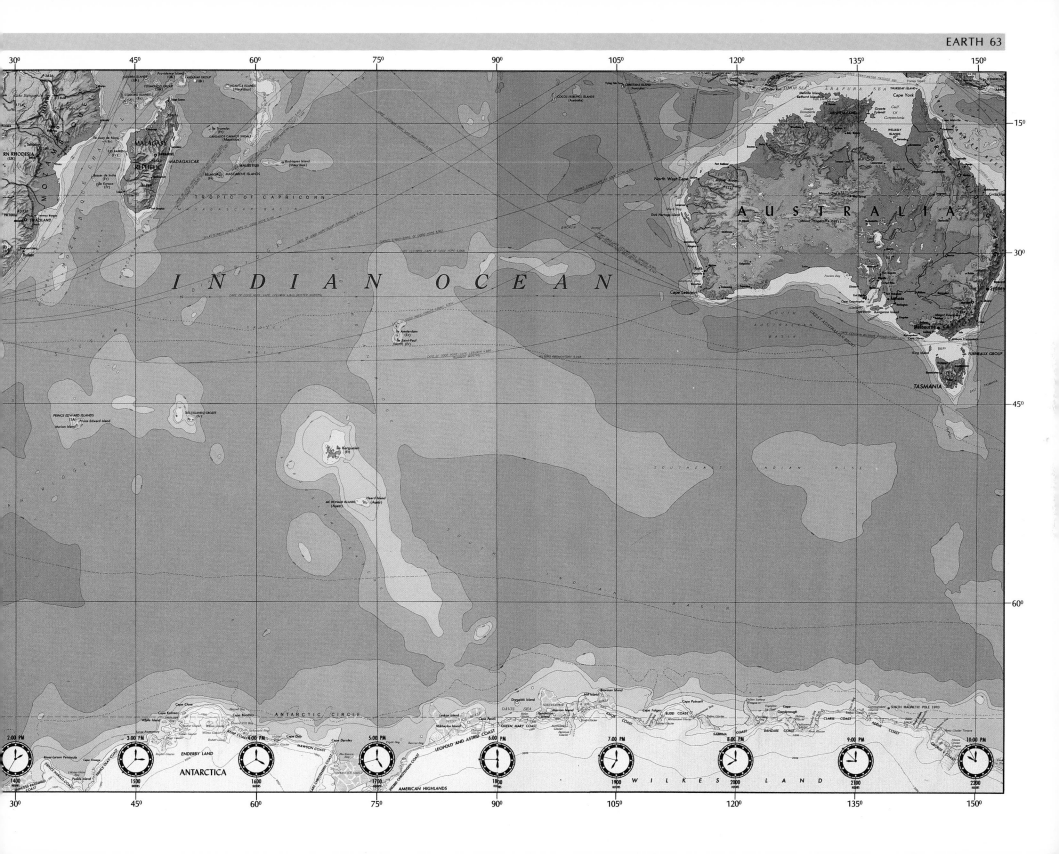

INDIAN OCEAN

AUSTRALIA

TASMANIA

ANTARCTICA

LIFE ON EARTH

Earth is the only planet in the Solar System, and indeed in the Universe, where life is *known* to exist. While it is possible that some form of life will be found to exist or to have once existed on another body in our Solar System, it is hard to imagine a world with anything to match the abundance of plant and animal life forms that have developed on the Earth over the past two billion years.

While life on Earth is thought to have originated at least two (and perhaps three) and a half billion years ago, its development is traced from a point 550 million years ago at the start of the

Cambrian Period. It is from the Cambrian Period that the oldest fossil evidence of life is found. The history of life on Earth after this point is measured in three geologic Eras, the *Paleozoic* (550-195 million years ago), the *Mesozoic* (195-60 million years ago) and the *Cenozoic,* which encompasses the past 60 million years. These Eras are in turn subdivided into Periods, of which the Cambrian reveals the earliest signs of other than the very most primitive life forms.

In *Precambrian* times, which are the oldest and largest division of geologic time, the earliest vestiges of life formed as a result of the amino acids present in the Earth's oceans. This plant-like bacteria used the process of photosynthesis, which plants still use, to turn sunlight into food. They used the water and carbon dioxide that was present, and in turn produced free oxygen. In the beginning, the free oxygen combined with metals (such as iron) to produce non-organic oxides (such as iron oxide or rust), but as the metals gradually became oxidized, oxygen gradually became a component of the Earth's atmosphere. As this happened, a new type of life form appeared that breathed oxygen just as plants breathed carbon dioxide. Thus it was that the two forks of the road of life on Earth, as we know it, began. These two forks—the oxygen-breathing animal kingdom and the carbon dioxide-breathing plant kingdom—became the two mutually interdependent life forms that have been perpetuated by their own symbiotic relationship for the past several million years.

By the dawn of the Cambrian Period the earliest and simplest single-celled life forms had come to share the Earth with a variety of multi-celled invertebrates. By the *Silurian* Period 195 million years later, the earliest vertebrate fishes had appeared in the sea, but it was not until the *Devonian* Period 50 million years later that life made the transition from the sea to the land. The *Carboniferous* Period (so-named because coal formation dates from this time) saw the development of insect life and the first, legged, amphibian creatures that could live on both land and sea. The Paleozoic ended with the end of the *Permian* Period 195 million years ago and set the stage for the ascendency of reptiles as the Earth's dominant animal form.

The Mesozoic Era, the Age of Reptiles, saw the advent of *dinosaurs* (literally *terrible lizards*), which were the largest creatures ever to walk the land surfaces of the Earth. While some of the dinosaurs were no larger than the reptiles that exist on Earth today, many varieties were much larger than any modern land animal. These included the fierce carnivore Tyrannosaurus, which stood 20 feet high on two legs, and the lazy herbivore, Brontosaurus, that might have weighed 50 tons and measured as much as 80 feet from nose to tail. The dinosaurs reached the high point of their development in the *Jurassic* and *Cretaceous* Periods—dominating life on Earth for 100 million years of the Mesozoic Era—before dying out suddenly and completely. It is not known why the dinosaurs became extinct in such a relatively short period of geologic time, but various theories have been advanced. A global cooling trend to which the dinosaurs could not adapt is probably part of the story, and recent evidence suggests that the Earth was struck by a huge extraterrestrial body such as a comet or asteroid in what is called the Terminal Cretaceous Event. Such an Event could have created a cataclysmic firestorm and increase in the Earth's global temperature, which would have killed a number of species. The Event would have

then resulted in a great deal of debris in the Earth's atmosphere for a number of years, which would have blocked the Sun's light and triggered such a global cooling trend as we have mentioned. The Cretaceous Period, however, also saw the development of bird life as well as the early stages of mammal development. These classes of animal survived the Event or series of Events that killed the dinosaurs, and went on to thrive in the present geologic era.

The Cenozoic Era, which began 60 million years ago, is called the Age of Mammals because during this time, mammals evolved from tiny and almost insignificant creatures to become the largest and dominant class on the Earth's land surfaces—and in the Earth's oceans as well. By the *Pleistocene* Period, one million years ago, many very large mammals such as mammoths, mastodons, saber-toothed tigers and large horses had developed. The Great Ice Age, which occurred 35,000 years ago, and which lasted 20,000 years—covering a quarter of the Earth with ice—saw the demise of many of the larger mammalian species and the rise of a peculiar new one. In the current Period, the *Holocene,* this new upright-walking species has come to be the Earth's dominant species. Beginning quietly in the Africa/Eurasia area, and gradually spread throughout the world, humans were the first animals to use tools, conduct widespread domestication of other species and to record their own history in a variety of spoken and written languages. The 'historic' segment of the Holocene Period began roughly 3000 years ago and roughly 20 years ago Humans became the first species to build tools to carry themselves away from the Earth to other bodies in the Solar System.

Facing page: This could well have been a scene in Precambrian times, when life on Earth consisted of monotremic animals such as those *above,* and before such life as the early dinosaurs (*at top*)—Plesiosaur (left) and Icthyosaur (right)—thrived.

THE EARTH'S GEOLOGIC ERAS AND THEIR PERIODS
(Data in parentheses indicate the number of years ago when the Period *ended*)

Azoic (Cosmic) Era	(2000 million)	Devonian Period	(305 million)
Archeozoic Era		Carboniferous Period	(220 million)
Keewatin Period	(1200 million)	Permian Period	(195 million)
Temiskaming Period	(1000 million)	**Mesozoic Era**	
Proterozoic Era		Triassic Period	(160 million)
Huronian Period	(800 million)	Jurassic Period	(125 million)
Algonquian Period	(550 million)	Cretaceous Period	(60 million)
Paleozoic Era		**Cenozoic Era**	
Cambrian Period	(480 million)	Tertiary Period	(1 million)
Ordovician Period	(395 million)	Pleistocene Period	(.015 million)
Silurian Period	(355 million)	Holocene Period	(present)

Above: Examples of animal life on Earth include the Giraffe (class mammalia) and Great White Heron (class aves). This page *from left* are a single-celled Protozoa, Monarch Butterfly (class insecta), Monte Verde Toad (class amphibia), Rattlesnake (class reptilia) and Brook Trout (class osteichthyes). All except Protozoa (phylum protozoa) and Monarch (phylum arthropoda) are from phylum chordata.

THE EARTH'S ANIMAL KINGDOM

The *phyla* are arranged from the simplest, and probably the oldest, to the most complex. Percentages indicate the number of species within each *phylum* as a proportion of the roughly one million animal species on Earth.

A. Phylum Protozoa (1.5 percent). The simplest animals, including single-celled animals, Protozoa are descendants of the original animals.
Class Infusoria (ciliate protozoa)
Class Mastigophora (flagellate protozoa)
Class Sarcodina (ameboid protozoa)
Class Sporozoa (spore-forming protozoa)

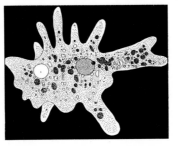

B. Phylum Porifera (.3 percent). Multi-celled stationary marine animals, Porifera are known familiarly as sponges.

C. Phylum Coelenterata (one percent). Invertebrate marine animals, Coelenterata include corals, sea anemones and jellyfish.

D. Phylum Ctenophora (.01 percent). Warm water marine animals, Ctenophora are similar to jellyfish, but lack stinging cells and use comb-like organs for locomotion.

E. Phylum Platyhelminthes (.7 percent). Simply described as flatworms, Platyhelminthes have a crude nervous system and can inhabit water, soil or the bodies of larger animals.
(1) *Class Cestoidea* (tapeworms)
(2) *Class Trematoda* (flukes)
(3) *Class Turbellaria* (free-living flat-worms)

F. Phylum Nemertinea (.05 percent). Simply described as ribbon worms, Nemertinea are the simplest animals to possess a complete digestive tract such as exists in higher animals.

G. Phylum Aschelminthes (.4 percent). Tiny aquatic animals, Aschelminthes in-clude roundworms and are all characterized by separate sexes.

H. Phylum Bryozoa or **Polyzoa** (.3 percent). Simply described as moss animals, Bryozoa are aquatic, with a u-shaped digestive tube and tentacles around their mouths.

I. Phylum Brachiopoda (.02 percent). Commonly known as bivalves or shellfish, Brachiopoda include clams, known from fossils to have represented a large percentage of the Earth's animal life in the early Paleozoic.

J. Phylum Chaetognatha (.003 percent). Found at various depths in the Earth's oceans, Chaetognatha are commonly known as arrow worms.

K. Phylum Annelida (.7 percent). Simply described as segmented or ringed worms, Annelida include the Atlantic palolo worm, as well as leeches and common earthworms.

L. Phylum Mollusca (4.5 percent). Soft-bodied, unsegmented animals, most Mollusks have a calcareous shell and a single 'foot.' They are common to both land and water.
Class Amphineura (chitons)
Class Cephalopoda (octopi and squids)
Class Gastropoda (snails and slugs)
Class Pelecypoda (oysters and mussels)
Class Scaphopoda (tooth shells)

M. Phylum Onychophora (.007 percent). Similar to Phylum Annelida, but having feet on each segment, Onychophora are land animals.

N. Phylum Arthropoda (84 percent). Known as joint-footed animals, Arthropods are characterized by a hard external skeleton and jointed legs. There are more species of Arthropods than all other Phyla combined.
Class Arachnida (spiders and scorpions). *Arachnids* are eight-legged Arthropods that constitute 35 percent of all animal species.
Class Crustacea (crust shelled animals). Constituting 25 percent of all animal species, Crustaceans include crayfish, lobsters and shrimp.

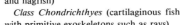

Class Diplopoda (millipedes)
Class Insecta (insects) are six-legged, air breathing Arthropods, usually with wings, and constitute nearly half of all of the Earth's animal species.
Class Myriapoda (centipedes)

O. Phylum Echinodermata (.5 percent). Described as spiny-skinned animals, Echinodermata include sea urchins, sea lilies, sea cucumbers and starfish.

P. Phylum Chordata (six percent). The most complex of all phyla, Chordata includes animals characterized by well developed nervous and digestive systems. Most are also vertebrates. The classes are organized below in order of their complexity, and the orders listed within each class are alphabetical.
Class Agratha (jawless animals with poorly developed back-bones, including lamprey and hagfish)
Class Chondrichthyes (cartilaginous fish with primitive exoskeletons such as rays)
Class Elasmobranchii (similar to the true fish of Class Osteichthyes, the Elasmobranchii, such as sharks, have a skeleton composed of cartilage rather than bone)
Class Osteichthyes (bony vertebrate fish that breathe by means of gills) The Osteichthyes include all true fishes such as bass, trout, salmon, sunfish, sturgeon, marlin, herring, flounder and pike.
Class Amphibia (Cold-blooded vertebrates who can live in both water and on land, breathing with lungs or gills) Amphibi-

an young are always born from eggs and live in water breathing with gills until they mature to adulthood. Existing Amphibian orders are as follows:
(a) Anura (frogs and toads)
(b) Caecilians (tropical legless amphibians)
(c) Caudata (salamanders and tailed amphibians)

Class Reptilia (cold-blooded, egg-laying animals that breathe by means of lungs) Except for snakes, Reptiles are all characterized by having four legs. Existing Reptilian orders are as follows:
(a) Crocodilia (alligators and crocodiles)
(b) Squamata (snakes and lizards)
(c) Rhynchocephalia (tuatara)
(d) Testudinata (turtles and tortoises)

Class Aves (One of only two warm-blooded animal classes, Aves, or birds, breathe with lungs, lay eggs and almost all are capable of flight.) Aves are divided into Subclass Archaeornithes and Subclass Neornithes. The former includes only a single species, the original bird type, the Jurassic Period's long-extinct Archaeopteryx, a feathered, but reptile-like toothed bird. Subclass Neornithes includes all other birds from other extinct species known only from fossils to all known existing types. The existing orders within this subclass are as follows:

(a) *Anseriformes* (ducks and geese)
(b) *Apodiformes* (hummingbirds and swifts)
(c) *Caprimulgiformes* (nightjars)
(d) *Charadriiformes* (gulls and ferus)
(e) *Ciconiiformes* (herons and storks)
(f) *Coliiformes* (moosebirds)
(g) *Columbiformes* (pigeons)
(h) *Coraciiformes* (kingfishers and hornbills)

(i) *Cuculiformes* (cuckoos)
(j) *Falconiforms* (eagles, falcons, hawks and vultures)
(k) *Gaviformes* (loons)
(l) *Galliforms* (grouse, pheasants and quails)
(m) *Groiformes* (cranes and rails)
(n) *Passeriformes* (blackbirds, bluebirds, canaries, cardinals, crows, jays, orioles, robins, thrushes, warblers and waxwings)
(o) *Pelecaniiformes* (pelicans)
(p) *Piciforms* (toucans and woodpeckers)
(q) *Phoenicopteriformes* (flamingos)
(r) *Podicipediformes* (grebes)
(s) *Procellariiformes* (albatrosses and petrels)
(t) *Psittaciformes* (parrots)
(u) *Spheniciformes* (penguins)
(v) *Strigiformes* (owls)
(w) *Struthioniformes* (kiwis, ostriches and rheas)
(x) *Tinamiformes* (tinamous)

Class Mammalia (Complex warm-blooded animals that breathe with lungs. With the exception of Order Monotremes (which lay eggs), all mammals give birth to living young. The orders of existing Mammalia are:

(a) *Artiodactyla* (even-toed ungulates such as cows, deer and pigs)
(b) *Carnivora* (bears, cats and dogs)
(c) *Cetacea* (dolphins and whales)

(d) *Chiroptera* (bats)
(e) *Dermoptera* (flying lemur)
(f) *Edentata* (anteaters and sloths)
(g) *Hyracoidea* (hyraxes)
(h) *Lagomorpha* (rabbits)
(i) *Marsupialia* (pouched animals such as kangaroos, koalas and opossums)
(j) *Monotremata* (platypus)
(k) *Perissodactyla* (odd-toed ungulates such as horses and rhinoceroses)
(l) *Pholidota* (scaly anteater)
(m) *Primates* (apes, humans and monkeys)
(n) *Proboscidea* (living and extinct elephants)
(o) *Rodentia* (beavers, rats and mice)
(p) *Sirenia* (dugongs and manatees)
(q) *Tubulidentata* (aardvark)

This page, left column from top: From phylum chordata's class aves: a Steller's Jay (order passeriforme) and an Eared Grebe (order podicipediforme). *Middle column and above:* From phylum chordata's class mammalia: a Nuttall's Cottontail (order lagomorpha), a Golden Mantled Ground Squirrel (order rodentia), a Jaguar (order carnivora) and a White-tailed Deer (order Artiodactyla).
Below: Examples of life from the Earth's plant kingdom include conifers from phylum strobilophyta (dark green) and deciduous trees from phylum cycadophyta (light green and yellow). Higher forms of plant life are generally green because of the presence of chlorophyl, which facilitates photosynthesis, the process by which green plants convert sunlight to chemical energy and synthesize organic compounds from inorganic compounds.
In cold climates, the green leaves of deciduous trees turn yellow or red and drop off in autumn as the trees become dormant for the winter.

THE EARTH'S PLANT KINGDOM

While botanists do not agree on a universal classification system, the outline below provides at least a glimpse of the variety of types of the Earth's plant life. They are listed by order of general complexity beginning with the simplest. The percentages indicate the number of species within each group as a proportion of the roughly 325,000 plant species on Earth.

A. Division Thallophyta (32 percent) Plants without stems, leaves or true roots
(1) *Phylum Schizophyta* (bacteria)
(2) *Phylum Evglenophyta* (mobile green—one-celled organisms that move by means of flagellum)
(3) *Phylum Cyanophyta* (blue-green algae)
(4) *Phylum Chrysophyta* (yellow-green algae)
(5) *Phylum Pyrrophyta* (dinoflagellate algae)
(6) *Phylum Chlorophyta* (green algae)
(7) *Phylum Rhodophyta* (red algae)
(8) *Phylum Phaeophyta* (brown algae)
(9) *Phylum Myxophyta* (slime molds)
(10) *Phylum Mycophyta* (fungi, including mushrooms)

B. Division Bryophyta (seven percent)
(1) *Phylum Bryophyta* (Mosses and Liverworts)

C. Division Tracheophyta (61 percent), Vascular plants
(1) *Phylum Psilophyta* (primitive ferns)
(2) *Phylum Lepidophyta* (club mosses)
(3) *Phylum Calamophyta* (horsetails)
(4) *Phylum Ptenophyta* (true ferns)

(5) *Phylum Cycadophyta* (cycads and true deciduous trees)
(6) *Phylum Strobilophyta* (conifers)
(7) *Phylum Gnetophyta* (aberrant gymnosperms)
(8) *Phylum anthophyta* (flowering plants)

THE EARTH'S MOON

Above: The mellow harvest Moon, once thought by Earth dwellers to be made of green cheese, reposes benignly above this northern hemispheric farm scene.

Though it is simply called 'the Moon,' the Earth's single satellite is mythologically associated with Luna (or Diana), the Roman goddess of the hunt, who was also their goddess of the Moon. The sixth largest moon in the Solar System, the Moon is closer in size to its mother planet than any other except Pluto's moon Charon. For this reason the Earth and the Moon (like Pluto and Charon as well) are occasionally described as being a *double planet.* While the larger planets have on the order of a thousand times the mass of their moons, the Earth has 81 times the mass of the Moon and four times its diameter.

Though it is the second brightest object in the sky, nearly all of the Moon's light is reflected from the Sun. Some tiny portion of the Moon's light could be called *Earthshine* as noticed during a crescent phase when the 'dark' part of the moon is slightly illuminated: this is light from the Sun reflected by the Earth to the Moon, which in turn reflects it back to the Earth. As perceived from Earth, the Moon appears to go through a series of phases depending upon its reflection of light from the Sun. These phases, which constitute the Lunar 'day,' go through a complete cycle every 29 days, 12 hours and 44 minutes. The cycle is also known as the *synodic* or *Lunar month,* as seen on Earth.

When the Moon is fully illuminated it is said to be 'full.' As the visible face of the Moon rotates away from the Sun it is said to be 'waning.' When exactly half the face of the Moon is illuminated, it is called a 'quarter Moon.' As it becomes less visible it is said to become a 'crescent Moon,' and when it becomes dark and the cycle is resumed, the Moon is said to be a 'new Moon.' From 'new,' the Moon waxes through the crescent phase to the quarter phase, and once again to full. The Sun always illuminates one half the Moon. Depending upon the relative angle between the Earth and Moon, we see portions of the sunlit side. At full Moon we see the entire sunlit side and at new Moon none of the sunlit side is facing the Earth.

The Moon's period of rotation is 27 days, 7 hours and 43 minutes—exactly the same as the period of its revolution around the Earth, so the same side always faces the Earth. Because of the Moon's slight wobbling, we are able to see slightly more than half of its surface from the Earth. The Moon's far side had been a mystery to mankind for centuries, and it was not until the Soviet Luna 3 spacecraft returned photographic images of the 'dark side of the Moon' in October of 1959 that actual detailed information of the Moon's 'other half' was revealed to mankind. Though the 41 percent of the Lunar surface that is never visible from the Earth is frequently referred to as the 'dark side,' it actually receives as much light from the Sun as the 'near' side. When the Moon is perceived as 'full' on Earth, the far side is in fact 'dark,' but as the near side gradually wanes, the Moon waxes on the far side, and vice versa.

The Moon's surface is characterized by rugged mountain ranges and by thousands of meteorite impact craters. In this sense it is very much like the planet Mercury. Unlike Mercury, however, the Moon has large open areas that are called *seas* (or in Latin, *maria*) because to the unaided eye, they appear darker than the surrounding terrain, and were once thought by Galileo to resemble seas. Almost entirely concentrated on the side facing the Earth, the maria cover 15 percent of the Lunar surface, and were probably once 'seas' of molten rock that flowed out of the Moon's interior. The gravitational effect of the Earth probably has a great deal to do with the fact that such features are concentrated on the Earthward side.

Data for the Earth's Moon

Diameter: 2159.89 miles (3476 km)
Distance from Earth: 252,698 miles (406,676 km)
Mass: 3.34×10^{22} lb (7.34×10^{22} kg)
Rotational period (Lunar day): 27.32 Earth days
Sidereal period (Lunar year): 27.32 Earth days
Inclination of rotational axis: 1.53°
Inclination to ecliptic plane (Earth = 0): 4° 58' to 5° 9'
Mean surface temperature: 0° F
Maximum surface temperature: 279° F
Minimum surface temperature: −273.2° F
Highest point on surface: The rim of the crater Newton
Largest surface feature: Mare Imbrium 384,400 sq mi (1,000,000 sq km)
Major atmospheric components: Traces of Hydrogen, Helium, Neon, Argon

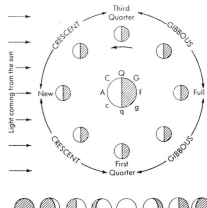

Left: This view of a near-full Moon was photographed by Apollo 13 on the return leg of that craft's Lunar voyage. Some of the successful Apollo manned Lunar mission program landing sites can be identified by using the manned mission table on page 73, and the Lunar maps on pages 75, 77 and 78-79.

Above: The phases of the Moon. Each lower circle marked A, c, q, g, etc shows how the Moon appears when viewed from a corresponding position (A, c, q, g, etc) on the Earth (represented by the central, upper circle).

Unlike the Earth, the Moon has neither magnetic poles nor a significant magnetic field, although rocks in the Lunar crust are weakly magnetized. Probably due to its low mass and density, the Moon never developed an atmosphere, although trace amounts of hydrogen and helium as well as hints of argon and neon were detected escaping from the Lunar surface in 1972 by the Apollo 17 crew.

The Apollo program studies revealed that the Lunar interior was quite active, with moonquakes being more common on the Moon than earthquakes are on the Earth, although moonquakes have not been recorded in excess of 2.0 on the Richter scale.

The Moon was formed about the same time as the Earth—4.6 billion years ago—and is composed of the same basic materials, but the early Earth/Moon relationship is unclear. One line of thought theorizes that the Moon was formed out of the Earth, either in a single piece that broke loose (perhaps from the Pacific Basin) or in the form of debris that was knocked loose in a collision with an asteroid, and which eventually congealed into the Lunar mass. Another theory holds that the Moon was a separate planet 'captured' by the Earth's gravitational field. A third notion has it that the Earth and Moon were formed in the same way and in the same place and time. Because the Earth was 81 times larger, the Moon became enslaved to its gravity.

Once in place, the Moon's geology evolved much like that of the Earth. Originally molten, the crust gradually cooled, leaving a molten core like that of the Earth. In the meantime the crust was also being bombarded by debris from the formation of the Solar System. In addition to smaller craters, huge basins were hammered into the surfaces of both the Earth and Moon. Some of the first basins to be formed on the Moon were Mare Fecunditatius and Mare Tranquilium, and they probably were formed 4.4 billion years ago. The last basins to be formed were the Mare Imbrium and Mare Orientale. Dating from 3.85 billion years ago, Mare Imbrium is the largest of the Lunar Seas and its origin concludes the *Preimbrian* Period of Lunar geologic evolution.

When the *Imbrian* Period began 3.85 billion years ago, the Lunar surface was probably pocked entirely and uniformly with impact craters. The semicircular mountain ranges found around the periphery of the maria are the only remnants of the enormous impacts that created them. During this period, however, intense interior heating resulted in vast flows of darker basalt from deep within the Moon. Part of the heating came from meteor impacts, and part from radioactive decay. These flows filled the huge basins, and the Lunar Seas briefly were seas—*of lava!*

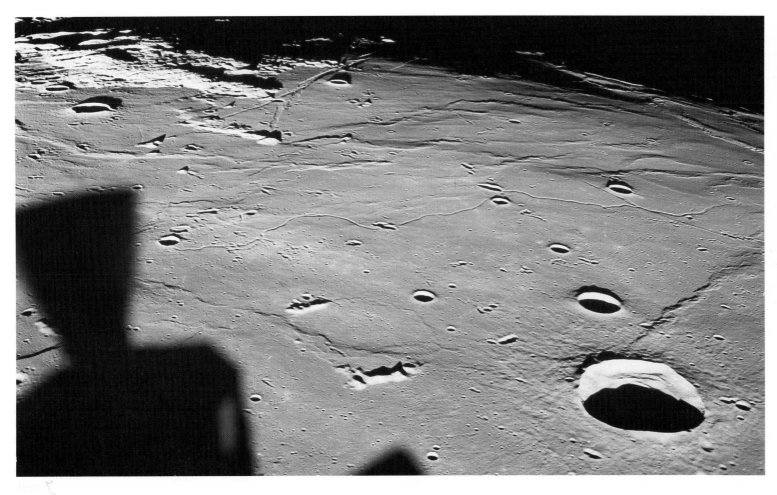

Facing page: The Apollo 12 Lunar Module *Intrepid,* with astronauts Charles Gordon, Jr and Alan Bean aboard, heads for its landing site in the Moon's Ocean of Storms. The large crater on the right side of this photo is Herschel.

Upper left: The approach to the Apollo landing site in the Sea of Tranquility is seen in this photo, taken by the Lunar Module itself (part of which is seen as the out-of-focus outline at lower left) as it prepared to descend. The crater Maakelyne is the large one at lower right and Hypatia Rille is at upper left center.

Above: A view from Apollo 14, looking south at the very intriguing Davey Crater Chain.

When the Moon cooled, and the lava flows ended 3.3 billion years ago, small scale volcanic activity continued for approximately 1.3 million years through what is called the *Eratosthenian* Period. During this period, interplanetary debris crashed into the Moon creating newer, smaller impact craters on the Lunar Seas themselves. One of the major craters now visible on the surface, Copernicus, was probably formed one billion years ago, marking the climax of the third period of Lunar geology. Since the formation of Copernicus, there has been very little geologic activity on the Moon. This fourth period, the present *Copernican* Period, has also been marked by very little in the way of impact crater formation, although the crater Tycho is thought to have been formed as recently as one million years ago, and the formation of the great crater Giordano Bruno is believed to have been witnessed from Earth in 1178 AD. In July 1972, a 2200 pound meteorite was recorded as having struck the Moon. Throughout the billion years of the Copernican Period, the Moon's surface has remained relatively unchanged because there is no air, no wind and no water to cause erosion of the type that has greatly altered the surfaces of such bodies as the Earth and Mars.

Being the first person to study the Lunar surface with a telescope, the great Italian astronomer Galileo Galilei (1564–1642) was also the first to publish a systematic Lunar Map (in 1610). In 1651 the Jesuit astronomer Father Joannes Riccioli (1598–1671) published a map of the Moon that is important, because the names he assigned to Lunar features are used by astronomers to this day. In the nineteenth century, photography came into play as a tool of Lunar cartography, and in 1935 the International Astronomical Union published the definitive map of the Lunar

near side. Early spacecraft confirmed the Earthbased observations and in 1959 the Soviet Luna 3 added the far side to what was known about Lunar geography.

The next major step in Lunar mapping came with the American Lunar Orbiter project. Between August 1966 and August 1967, five Lunar Orbiter spacecraft conducted high-resolution photography of 99 percent of the Lunar near side and 80 percent of the Lunar far side. Between June 1966 and February 1968, a series of five (out of seven launched) American Surveyor spacecraft were successfully landed on the Lunar surface to conduct remote sampling of Lunar surface material. The Lunar Orbiter

Manned Lunar Orbital Missions

Spacecraft	Crew	Launch	Lunar Encounter	Number of Orbits
Apollo 8	Frank Borman James Lovell William Anders	21 Oct 68	24 Dec 68	10
Apollo 10	Eugene Cernan Thomas Stafford John Young	18 May 69	21 May 69	31
Apollo 13	Fred Haise James Lovell Jack Swigert	11 Apr 70	14 Apr 70	1/2

(intended to be a Lunar landing but system malfunctions resulted in mission abort)

Successful Unmanned Expeditions from Earth to the Moon

Spacecraft	Country of Origin	Launch	Date of Lunar Encounter	Results
Pioneer 1	USA	11 Oct 58		Failed, but sent 43 hours data
Pioneer 3	USA	6 Dec 58		Failed, but sent radiation data
Luna 1	USSR	2 Jan 59	4 Jan 59	3717 mile approach, then went into solar orbit
Pioneer 4	USA	3 Mar 59	4 Mar 59	37,282
Luna 2	USSR	12 Sep 59	13 Sep 59	Impacted on surface
Luna 3	USSR	4 Oct 59	10 Oct 59	Orbited at 3844 miles
Ranger 7	USA	28 Jul 64	31 Jul 64	Returned 4308 pictures, then impacted in Sea of Clouds
Ranger 8	USA	17 Feb 65	20 Feb 65	Landed, relayed 5814 pictures, impacted in Crater Alphonsus
Zond 3	USSR	18 Jul 65	20 Jul 65	5715 mile approach
Luna 9	USSR	31 Jan 66	3 Feb 66	Landed; sent back first TV transmission from service
Luna 10	USSR	31 Mar 66	3 Apr 66	217 mile approach; sent back data for 2 months
Surveyor 1	USA	30 May 66	4 Jun 66	World's first fully controlled soft landing
Lunar Orbiter 1	USA	10 Aug 66	29 Oct 66	Deliberately impacted
Luna 11	USSR	24 Aug 66	29 Aug 66	99 mile approach; orbited until 1 Oct 66
Luna 12	USSR	22 Oct 66	25 Oct 66	Orbited until 19 Jan 67
Lunar Orbiter 2	USA	6 Nov 66	19 Nov 66	Lunar satellite

Spacecraft	Country of Origin	Launch	Date of Lunar Encounter	Results
Luna 13	USSR	21 Dec 66	24 Dec 66	Landed; sent back panoramic views, measured soil density and surface radiation for 6 days
Lunar Orbiter 3	USA	4 Feb 67	15 Feb 67	Lunar satellite; deliberately impacted
Surveyor 3	USA	17 Apr 67	20 Apr 67	Landed; made first excavation and bearing test on an extraterrestrial body
Lunar Orbiter 4	USA	4 May 67	8 May 67	Provided first pictures of Lunar south pole; deliberately impacted
Explorer 35	USA	19 Jul 67	22 July 67	Placed in Lunar orbit; Studied Moon's magnetic field and radiation belt
Lunar Orbiter 5	USA	1 Aug 67	31 Jan 68	Lunar satellite; controlled impact
Surveyor 5	USA	8 Sep 67	10 Sep 67	Landed in Sea of Tranquility; did first on-site chemical soil analysis
Surveyor 6	USA	7 Nov 67	10 Nov 67	Landed in Sinus Medii; performed first takeoff from Lunar surface
Luna 14	USSR	7 Apr 68	10 Apr 68	99 mile approach
Zond 5	USSR	15 Sep 68	18 Sep 68	1209; returned to Earth 21 Sep 68

Spacecraft	Country of Origin	Launch	Date of Lunar Encounter	Results
Zond 6	USSR	10 Nov 68	14 Nov 68	1500 mile approach; returned to Earth 17 Nov 68
Luna 15	USSR	13 July 69	21 Jul 69	Crashed on Moon after 52 orbits
Zond 7	USSR	8 Aug 69	11 Aug 69	1240 mile approach; returned to Earth 14 Feb 70
Luna 16	USSR	12 Sep 70	20 Sep 70	Landed; recovered 24 Sep 70
Zond 8	USSR	20 Sep 70		Returned to Earth 17 Oct 70
Luna 17	USSR	10 Nov 70	17 Nov 70	Landed: carried first Moon robot, Lunokhod 1
Luna 18	USSR	2 Sep 71	11 Sep 71	Crashed on Moon after 54 orbits
Luna 19	USSR	28 Sep 71	3 Oct 71	87 mile approach; made over 4000 Lunar orbits
Luna 20	USSR	14 Feb 72	21 Feb 72	Landed and returned to Earth 25 Feb 72
Luna 21	USSR	8 Jan 73	16 Jan 73	Carried Lunokhod 2 to the Moon
Explorer 49	USA	10 Jun 73	15 Jun 73	688 mile approach; transmitted through 1975
Luna 22	USSR	29 May 74	9 Jun 74	151 × 15.5 mile Lunar orbit
Luna 23	USSR	28 Oct 74	6 Nov 74	Landed, but equipment was damaged and program abandoned after 3 days
Luna 24	USSR	9 Aug 76	17 Aug 76	Landed and returned to Earth 22 Aug 76

and Surveyor projects increased our knowledge of the Moon far beyond what had been known before, but they were soon to be overshadowed by the project for which they had been designed to pave the way.

In 1968, the United States began the Apollo project, a series of spaceflights during which the Moon became the first body in the Solar System beyond Earth to be explored firsthand by human beings.

The Moon was surveyed by human beings from Lunar orbit for the first time by means of two circumlunar manned flights in December 1968 and May 1969, which began the operational phase of the Apollo program. In July 1969, the Apollo 11 spacecraft became the first vehicle to land human beings on the Moon. This initial landing was followed by five others between November 1969 and December 1972 (a seventh mission was aborted because of hardware failure in April 1970).

During the Apollo program, 12 American astronauts conducted detailed surveys of the Lunar surface and seismic studies of the Lunar interior. The Apollo program completed detailed mapping of the Moon and provided a wealth of information about its composition and its geologic history.

Facing page: Astronaut John W Young, commander of the Apollo 16 Lunar landing mission, leaps from the Moon's surface as he salutes the US flag at the crater Descartes. The Lunar Module *Orion* is on the left and the Lunar Roving Vehicle is parked beside the LM. Stone Mountain dominates the background in this scene.

Above: Geologist-Astronaut Harrison H Schmitt is shown here working at the Lunar Rover, near the Apollo 17 LM base in the Sea of Serenity. Shorty Crater is to the right and the peak in the center background is Family Mountain.

Manned Lunar Landing Missions

Spacecraft	Crew	Launch	Lunar Landing	Landing Site	Duration on Surface	Lunar Material Returned
Apollo 11	Edwin 'Buzz' Aldrin* Neil Armstrong* Michael Collins	16 Jul 69	20 Jul 69	Sea of Tranquility	21 hrs, 32 min	48.07 lbs
Apollo 12	Alan Bean* Charles Conrad* Richard Gordon	14 Nov 69	19 Nov 69	Ocean of Storms	31 hrs, 30 min	75.14 lbs
Apollo 14	Edgar Mitchell* Stuart Roosa Alan Shepard*	31 Jan 71	5 Feb 71	Frau Mauro crater	22 hrs, 30 min	94 lbs
Apollo 15	James Irwin* David Scott* Alfred Worden	26 Jul 71	30 Jul 71	Hadley-Appenine site	66 hrs, 55 min	172.4 lbs
Apollo 16	Charles Duke* Thomas (Ken) Mattingly John Young*	16 Apr 72	30 Jul 71	Cayley Plains of Descartes	71 hrs, 2 min	209 lbs
Apollo 17	Eugene Cernan* Ronald Evans Dr Harrison (Jack) Schmitt*	7 Dec 72	11 Dec 72	Sea of Serenity	74 hrs, 59 min	249.7 lbs

*Two members of each crew actually walked on the Lunar surface, the third remained in Lunar orbit

Northern Hemisphere (Near Side) Geologic Map

At left is a photomicrograph of a thin section of an Apollo 17 Lunar rock sample. Obtained from the vicinity of Shorty Crater, this sample is a porphyritic, coarse-grained basalt, and was part of a 249.7 pound payload of Lunar samples brought back to Earth by the Apollo 17 crew.

Below: Crater Kepler and vicinity. This photo was taken from the Apollo 12 Orbital Module in November of 1969.

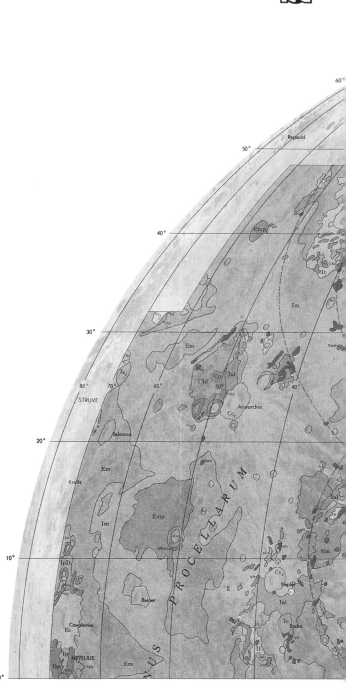

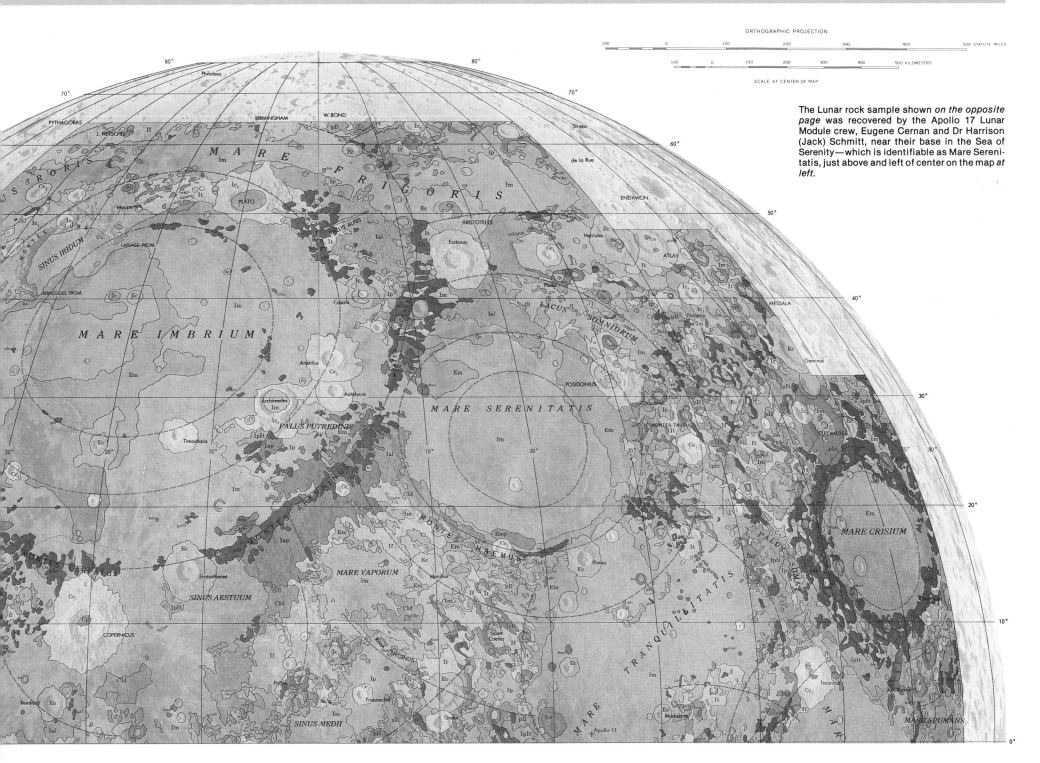

ORTHOGRAPHIC PROJECTION

100   0   100   200   300   400   500 STATUTE MILES

100   0   100   200   300   400   500 KILOMETERS

SCALE AT CENTER OF MAP

The Lunar rock sample shown *on the opposite page* was recovered by the Apollo 17 Lunar Module crew, Eugene Cernan and Dr Harrison (Jack) Schmitt, near their base in the Sea of Serenity—which is identifiable as Mare Serenitatis, just above and left of center on the map *at left.*

MARE FRIGORIS

MARE IMBRIUM

MARE SERENITATIS

LACUS SOMNIORUM

SINUS IRIDUM

PALUS PUTREDINIS

MONTES APENNINUS

MONTES HAEMUS

MARE VAPORUM

SINUS AESTUUM

MONTES CARPATUS

SINUS MEDII

MARE TRANQUILLITATIS

MARE CRISIUM

PALUS SOMNII

MARE SPUMANS

COPERNICUS

PLATO

ARISTOTELES

ENDYMION

ATLAS

MESSALA

CLEOMEDES

POSIDONIUS

MONTES TAURUS

PYTHAGORAS

PHILOLAUS

BIRMINGHAM

W. BOND

Strabo

de la Rue

Hercules

Eudoxus

Cassini

Archimedes

Autolycus

Timocharis

Eratosthenes

Manilius

Triesnecker

Godin

Plinius

Menelaus

Julius Caesar

Maskelyne

Apollo 11

RIMA HYGINUS

Reinhold

Copernicus

Geminus

Franklin

+ Apollo 11

Southern Hemisphere (Near Side) Geologic Map

DESCRIPTION OF MAP UNITS

CRATER MATERIALS
(Only craters 20 km or more in diameter are mapped)

Cc MATERIAL OF VERY SHARP-RIMMED RAYED CRATERS

Ec MATERIAL OF SHARP-RIMMED CRATERS

Ic₂ MATERIAL OF SINGLE CRATERS —Orientale basin age or younger

Ic₁ MATERIAL OF SINGLE CRATERS—Older than Orientale basin and younger than Imbrium basin

Ic MATERIAL OF IMBRIAN CRATERS, UNDIVIDED

Ico MATERIAL OF ELONGATE CLUSTERS OF CRATERS APPROXIMATELY RADIAL TO ORIENTALE BASIN—*Interpretation:* secondary craters of the Orientale basin

Icc MATERIAL OF ELONGATE AND IRREGULAR CLUSTERS OF CRATERS— Some not obviously related to any particular Imbrian basin or crater, others secondary craters of adjacent Imbrian crater to which they are approximately radial

Ifc FURROWED, RAISED CRATER FLOOR OR FILL MATERIAL—Most are convex upward. Texture equally sharp in Nectarian and Imbrian craters. *Interpretation:* Postimpact filling, probably lava

Nc MATERIAL OF SUBDUED CRATERS—Older than Imbrium basin and younger than Nectaris basin

Ncc MATERIAL OF ELONGATE CLUSTERS OF SUBDUED CRATERS—Craters approximately the same size. *Interpretation:* secondary craters of the Nectarian basin to which they are adjacent and (or) approximately radial

pNc MATERIAL OF SUBDUED TO VERY SUBDUED CRATERS—Older than Nectaris basin

BASIN MATERIALS
(Multiringed circular structures 300 km or more in diameter as measured across most prominent ring)

Nb MATERIAL OF RAISED RIMS AND SLUMPED WALLS OF BASINS—Primarily the outermost ring. *Interpretation:* disrupted bedrock largely covered by ejecta

Nbl LINEATED MATERIAL SURROUNDING BASINS—Linear elements approximately radial to basin; locally only weakly developed. *Interpretation:* basin ejecta deposited ballistically and as massive flows

NpNbm MATERIAL OF NECTARIAN BASIN MASSIFS—Massive mountain blocks, mainly form part of outermost basin rings; inner ring in Korolev. *Interpretation:* uplifted and structurally complex blocks of prebasin bedrock; may be covered by basin ejecta

NpNbr MATERIAL OF RUGGED MOUNTAINS AND MOUNTAIN SEGMENTS OF NECTARIAN BASINS—Smaller than massifs (unit NpNbm); mainly form inner rings. *Interpretation:* uplifted and complexly faulted prebasin bedrock; may be covered by basin ejecta

pNb PRE-NECTARIAN BASIN MATERIAL, UNDIVIDED—Subdued, eroded mountain rings and arcuate segments of rings

pNbm MATERIAL OF PRE-NECTARIAN BASIN MASSIFS—Relatively massive single mountain blocks or part of continuous ring. *Interpretation:* same as unit NpNbm

pNbr RUGGED MATERIALS OF PRE-NECTARIAN BASINS—Discontinuous blocky mountains forming arcuate ring segments; smaller than massifs (unit pNbm). *Interpretation:* same as unit NpNbr

OTHER MATERIALS

Im MARE MATERIALS OF DARK PLAINS—*Interpretation:* basaltic lavas, by analogy with returned Apollo samples. STIPPLED PATTERN: light streaks and swirls in Mare Ingenii. *Interpretation:* Surficial markings of uncertain origin, probably not related to underlying mare rocks

Ig MATERIAL OF GROOVES AND MOUNDS—Covers craters and other terrae of pre-Nectarian through Imbrian age. Craters have mainly radial grooves on rim and walls; some mounds. Level terra has mounds and grooves. Particularly well developed around Mare Ingenii and crater Van de Graaff. *Interpretation:* origin uncertain; general area antipodal to Imbrium basin; therefore could be depositional site of Imbrium ejecta that traveled around the Moon, or mass-wasting caused by Imbrium seismic shaking; alternatively, may be some unidentified local phenomenon unique to this area

Ip SMOOTH LIGHT PLAINS—Generally higher density of craters than on maria. *Interpretation:* may be related to formation of an Imbrian basin

INp LIGHT PLAINS—Higher density of craters than unit Ip. *Interpretation:* may be related to various Imbrian and/or Nectarian basins

Np HIGHLY CRATERED LIGHT PLAINS—*Interpretation:* may be related to Nectarian basins

It RELATIVELY FRESH-APPEARING, IRREGULAR TERRA—Low relief; low density of superposed craters. *Interpretation:* probably a complex mixture of local erosional debris and crater and basin ejecta

Nt ROLLING TERRA—Moderately high density of craters, particularly craters of diameter less than 20 km. *Interpretation:* same as unit It

NpNt IRREGULAR TERRA—Covers large areas and has high density of craters larger than 20 km wide. *Interpretation:* same as unit It

pNt CRATERED TERRA—High density of arcuate low hills or crater segments, and pre-Nectarian craters. *Interpretation:* same as unit It except contains erosional remnants of pre-Nectarian craters

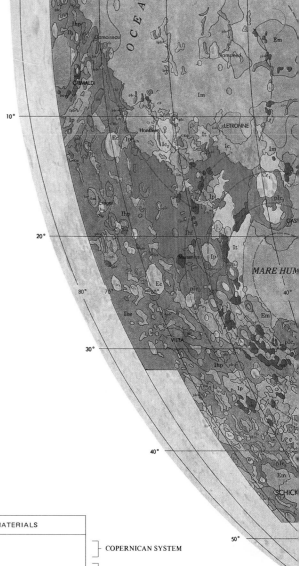

CORRELATION OF MAP UNITS

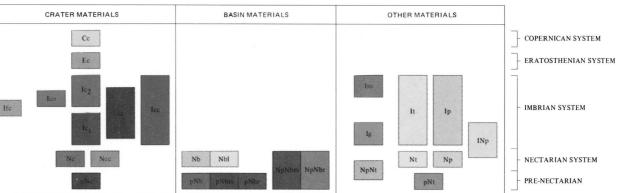

CRATER MATERIALS	BASIN MATERIALS	OTHER MATERIALS	
Cc			COPERNICAN SYSTEM
Ec			ERATOSTHENIAN SYSTEM
Ico Ic₂ Ifc Ic Icc Ic₁	Nb Nbl NpNbm NpNbr	Im Ig It Ip INp	IMBRIAN SYSTEM
Nc Ncc	pNb pNbm pNbr	Nt Np	NECTARIAN SYSTEM
pNc		NpNt pNt	PRE-NECTARIAN

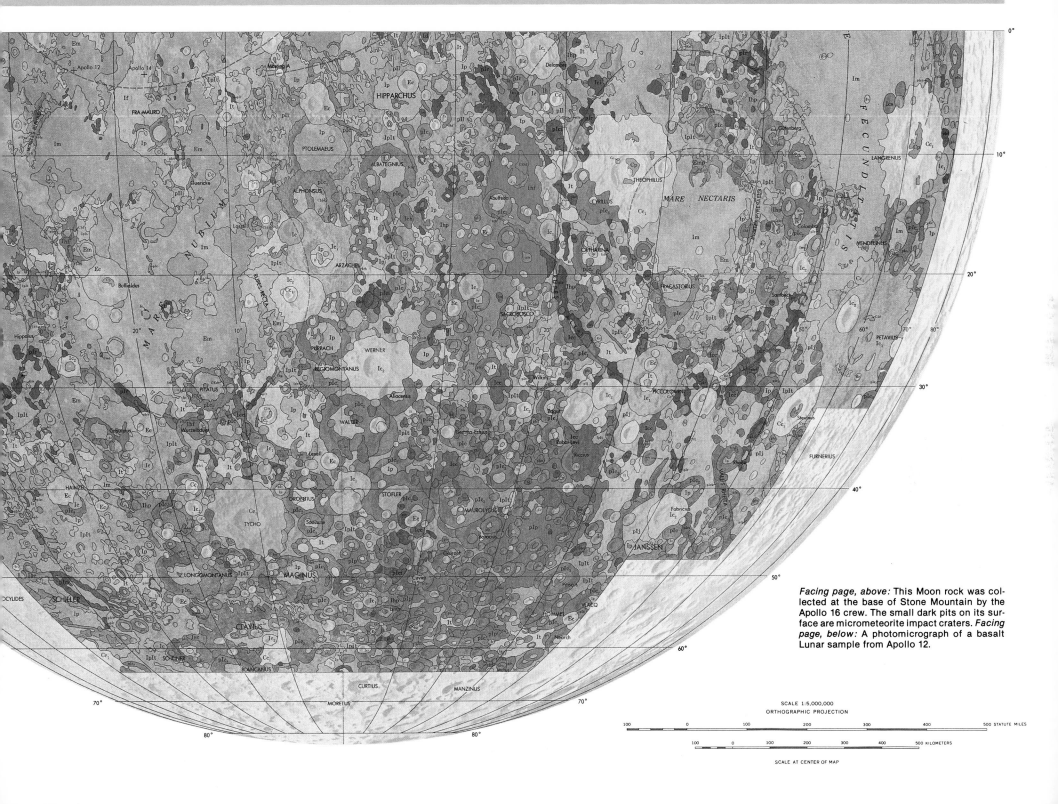

Facing page, above: This Moon rock was collected at the base of Stone Mountain by the Apollo 16 crew. The small dark pits on its surface are micrometeorite impact craters. *Facing page, below:* A photomicrograph of a basalt Lunar sample from Apollo 12.

SCALE 1:5,000,000
ORTHOGRAPHIC PROJECTION

100   0   100   200   300   400   500 STATUTE MILES

100   0   100   200   300   400   500 KILOMETERS

SCALE AT CENTER OF MAP

Mercator Projection

Systematic regional mapping shows the Moon to be a primitive body without the orogenic belts, mobile plates, oceanic ridges and widespread, water borne sedimentary deposits characteristic of the Earth's crust. Both volcanoes and meteors have shaped the Moon—therefore, though not as intensively reworked as the Earth, the Moon does exhibit a geologically heterogeneous surface, and even rocks which are relatively young on the Lunar time scale are as old or older than any presently dated terrestrial rocks; the geologic record on the Moon appears to complement that of the Earth by covering the period for which the record is missing on Earth.

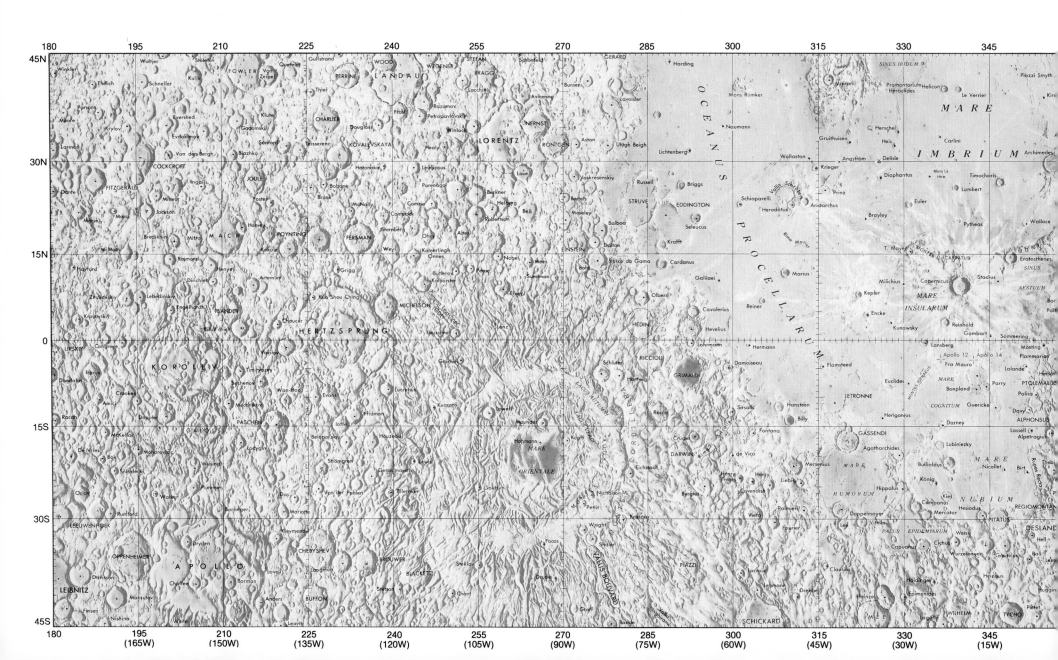

At left: The area covered by this photo is approximately 20 miles to a side, and the photographed area is located at 3° South, 160° West on the Moon's far side. The Apollo 8 Lunar orbital mission crew captured this view with a telephoto lens equipped camera.

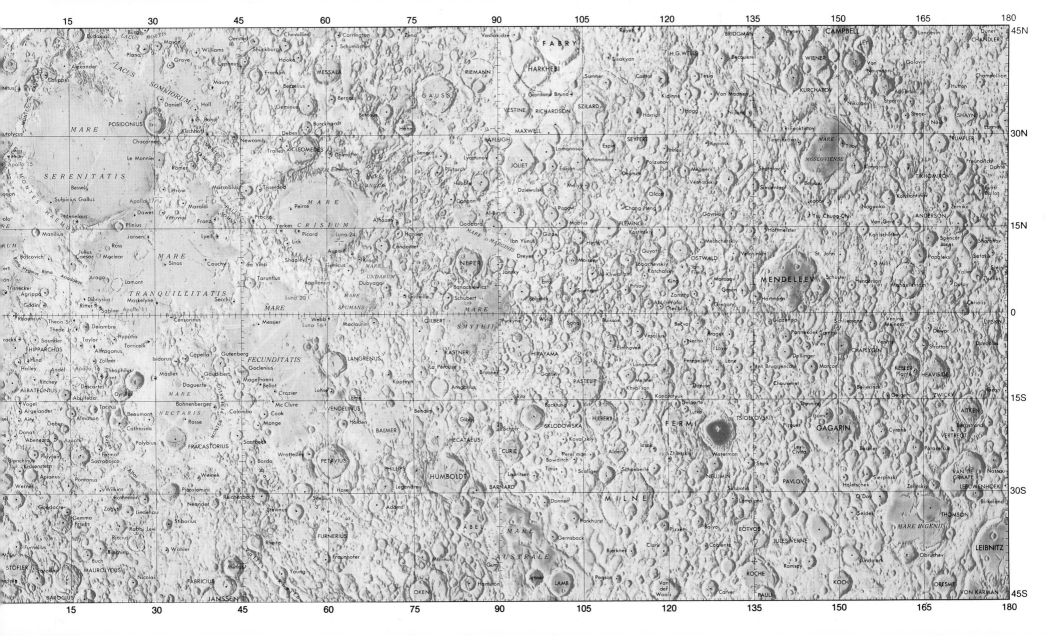

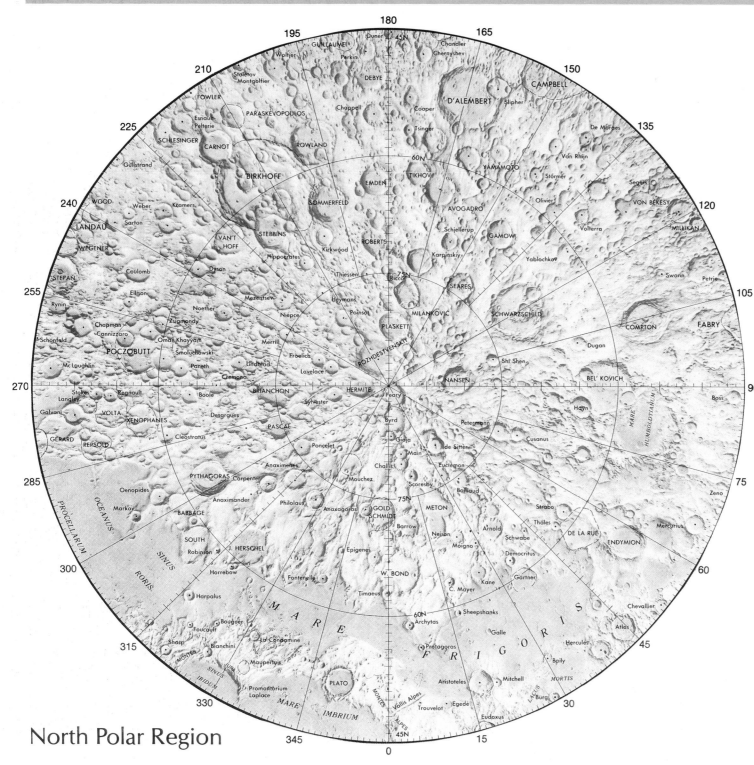

North Polar Region

Above: The Apollo 15 Command and Service Modules are shown here as they orbited above the Moon's northern region in July of 1971. This photo was taken from the Apollo 15 Landing Module, which was en route to the Lunar surface for an 18 hour, 37 minute stay.

Below: An Apollo 10 westward view across Apollo Landing Site 3 in the Central Bay. The prominent crater Bruce, at the bottom of this photo, is 3.7 miles wide.

Apollo 17 Geologist-Astronaut Harrison H Schmitt is shown *above,* as he conducted experiments near a huge, split Lunar boulder near the Taurus-Littrow landing site (in the Sea of Serenity), in December of 1972.

Below: The crater Langrenus as photographed from an altitude of nearly 350 miles, by Apollo 10 in December of 1968. Langrenus is approximately 85 miles in diameter.

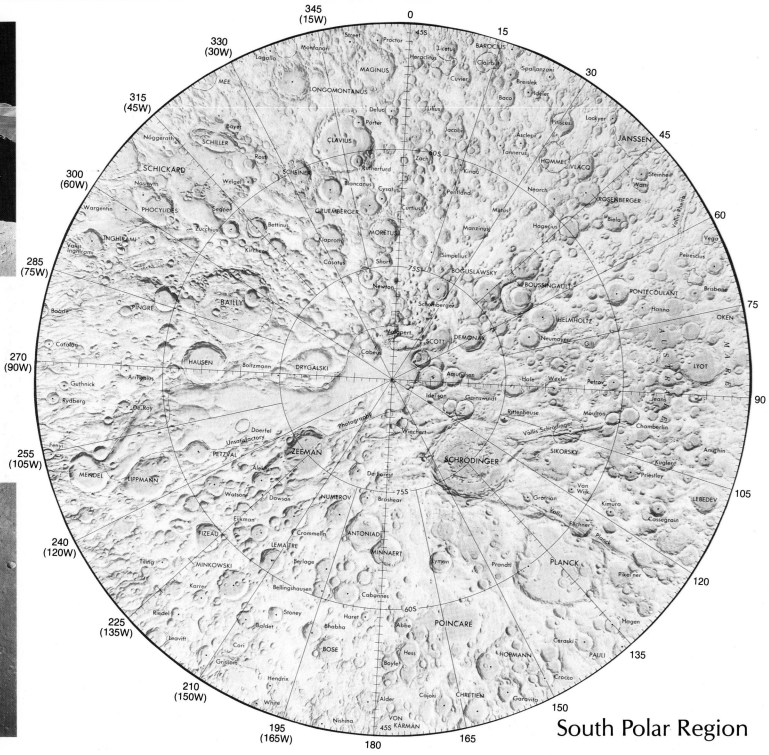

345
(15W)

330
(30W)

315
(45W)

300
(60W)

285
(75W)

270
(90W)

255
(105W)

240
(120W)

225
(135W)

210
(150W)

195
(165W)

180

165

150

135

120

105

90

75

60

45

30

15

0

Street Proctor Licetus BAROCIUS
Montanari Heraclitus Clairaut Spallanzani
Lagolla Cuvier Breislek
MAGINUS Jacobi Baco Ideler
LONGOMONTANUS Lilius Asclepi
MEE Deluc Porter Tannerus Pitiscus Lockyer
Nöggerath Bayer CLAVIUS Zach Kinau Nearch JANSSEN Steinheil
SCHILLER Rost Rutherfurd 60S HOMMEL Watt VLACQ
SCHICKARD Weigel SCHEINER Blancanus Cysatus Pentland Matus Manzinus Hagecius ROSENBERGER Biela
Nasmyth Segner GRUEMBERGER Curtius Simpelius Neander Vega
Wargentin Zucchius Bettinus Klaproth Short 75S BOGUSLAWSKY Peirescius
PHOCYLIDES Kircher MORETUS Casatus Newton Schomberger BOUSSINGAULT Brisbane
INGHIRAMI Bailly Malapert DEMONAX HELMHOLTZ PONTÉCOULANT OKEN
Vallis Inghirami Baade PINGRÉ Cabeus SCOTT Neumayer Hanno Gill
Catalan HAUSEN Boltzmann DRYGALSKI Amundsen Hale Wexler Petrov LYOT
Arrhenius Idel son Gainswindt Rittenhouse Moulton Chamberlin
Guthnick De Roy Photography Wiechert Vallis Schrödinger Jeans
Rydberg Doertel Unsatisfactory SCHRÖDINGER SIKORSKY Anuchin
Fenyi PETZVAL ZEEMAN De Forest Grafrian Van Wijk Kugler
MENDEL Alekhin Kimura Priestley
LIPPMANN Watson NUMEROV 75S Féchner LEBEDEV
Dawson Brashear Vallis Cassegrain
Eijkman Planck
FIZEAU Crommelin ANTONIADI Lyman Prandtl PLANCK
Tiling LEMAITRE MINNAERT Pikelner
MINKOWSKI Berlage 120
Karrer Bellingshausen Cabannes Hagen
Riedel Stoney Haret Abbe POINCARÉ Ceraski PAULI
Baldet Bhabha HOFMANN Crocco
Leavitt Cori Hess Boyle Hendrix BOSE Cajori CHRÉTIEN Garavita
White Nishina Alder VON KÁRMÁN 150
60S 45S

South Polar Region

MARS

Known as the 'red planet' because of its distinctive iron oxide coloration, Mars reminded early observers of a distant bloody battlefield and the Romans named it Mars after their god of war. Because of its perceived similarity to Earth, Mars has interested and intrigued Earthbound observers for centuries, and that is certainly still the case in the twentieth century.

Venus may be the Earth's near twin in terms of size, but Mars has more specific characteristics in common with the Earth than does any other planet. The Martian year lasts 23 Earth months, but the Martian day is only 41 minutes longer than the Earth day and its inclination of 24 degrees to its rotational axis is very close to the Earth's 23.4 degrees. As a result, Mars has four seasons that parallel those that we experience on Earth. The Martian summer is relatively warmer than its winter, which is characterized in the temperate zones by occasional light snow or frost. Mars is the only planet in the Solar System besides the Earth that has polar ice caps, and these can be seen to expand and recede seasonally like those on Earth. The south polar cap is composed of both water, ice and carbon dioxide ice (dry ice). The north polar cap is composed only of water ice with a residual carbon dioxide snow cover that evaporates in summer.

Mars is a good deal smaller and less dense than Earth. Because of this, Mars has less gravity and a much thinner atmosphere than Earth. It is also colder than the Earth, with temperatures in its polar regions rarely rising above the −200 degrees Fahrenheit level. The midsummer temperature near the Martian equator can, however, reach a comfortable 80 degrees Fahrenheit, closer to an Earthly temperature range than can be found anywhere else in the Solar System.

The characteristic rust-red Martian surface is indicative of the global high concentration of iron oxide in its soil and rocks. The surface of Mars is covered by the types of volcanic and impact craters that are found on Mercury or on the Earth's Moon, but there are also vast lightly cratered plains (or *planitia*), particularly in the northern hemisphere. While there are no currently active volcanos on Mars, those dormant volcanoes which do exist are the tallest volcanoes yet discovered in the Solar System.

The dormant shield-type volcano, Olympus Mons (Mount Olympus), is the tallest mountain on Mars, soaring 15.5 miles above the mean radius or standard elevation of the Martian surface. Mount Olympus stands 10 miles higher than the Earth's Mount Everest and encompasses more than 50 times the volume of Hawaii's Mauna Loa, the largest shield volcano on Earth. Southeast of Mount Olympus across the plain of the Martian Tharsis Region stands a neat row of three other large and important shield volcanic mountains—Arsia, Pavonis and Ascraeus. Ancient lava flows from the Martian volcanos helped to create Mars' vast open plains.

Another feature common to other bodies in the inner Solar System are fault rifts and seismic fracture zones. One such large fracture is Valles Marineris (Mariner Valley), a huge canyon stretching east by southeast from the Tharsis Region across the Martian equator. More than 3000 miles in length, Valles Marineris is the largest single surface feature on the planet. More than four times deeper than the Grand Canyon on Earth, it is a network of roughly parallel rift canyons with an overall width of up to 400 miles and a main canyon 125 miles wide at its widest.

In addition to craters and rift canyons, the Martian surface is

Prima Martis facies

M.

Occ. ——— Orie

S.

Primæ faciei
Succeſſiua conuersio

Above: Note the four ice caps represented on this early map of Mars, which may make an interesting comparison with the Mars photo on page 85, and with the various maps that follow in this section.

Facing page: This computer enhanced photo of Mars was taken by Viking Orbiter 2 in August of 1976. Visible features are, from left to right, ice cloud plumes on the western flank of Ascreaus Mons, one of the giant Martian volcanoes; the great rift canyon, Valles Marineris; and the large, frosty crater basin called Argyre.

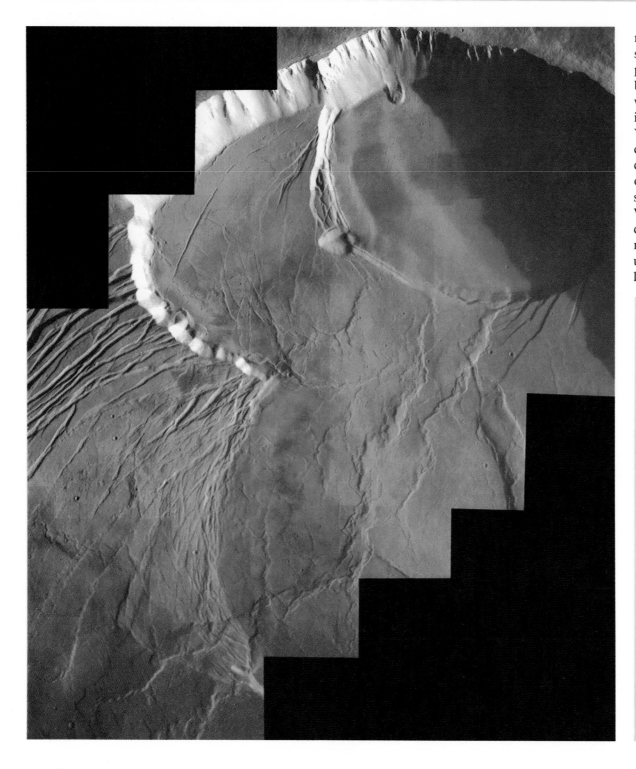

marked by vast networks of dry riverbeds, huge channels cut by streams of running water. These features defy explanation on a planet whose water is frozen in polar ice caps or trapped deep below the surface in subterranean permafrost. There is no liquid water visible at any place on the surface of the red planet, nor is Mars' atmospheric pressure high enough to permit it to exist. Yet the riverbeds bear silent witness to the fact that a great deal of water may have once flowed there. The channels can be divided into three categories—run-off channels with networks of tributaries, outflow channels that flowed from underground sources, and wide flat-floored fretted channels or flood plains. While there is no clue as to where the water disappeared, the evidence suggests large scale and widespread flooding. While some riverbeds cut across craters, there are other craters superimposed upon them, thus indicating that the flooding took place over a long period of time.

Data for Mars

Diameter: 4212.28 miles (6794 km)
Distance from Sun: 154,380,000 miles (249,000,000 km) at aphelion
127,720,000 miles (206,000,000 km) at perihelion
Mass: 2.9178×10^{23} lb (6.4191×10^{23} kg)
Rotational period (Martian day): 24.62 Earth hours
Sidereal period (Martian year): 687 Earth days
Eccentricity: 0.093
Inclination of rotational axis: 23.98°
Inclination to ecliptic plane (Earth=0): 1.85°
Albedo (100% reflection of light=1): .16
Maximum surface temperature: 78.8° F
Minimum surface temperature: −193.3° F
Highest point on surface: Mt Olympus (Olympus Mons)
Largest surface feature: Valles Marinaris (over 3000 miles long & 434.9 miles at the widest point)
Major atmospheric components:
Carbon dioxide (95.32%)   Oxygen (.13%)
Nitrogen (2.7%)   Carbon monoxide (.07%)
Argon (1.6%)   Water vapor (.03%)
Other atmospheric components: Neon, Krypton, Xenon, Ozone

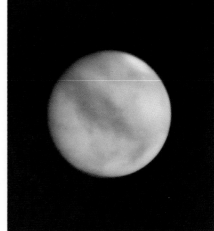

Beneath the Martian surface is a rigid crust 31 miles thick which probably contains water ice permafrost. The Martian mantle, composed of basalt rock, is roughly 125 miles thick. Beneath the mantle is a partially-molten transition zone leading to a formerly molten core that is between 800 and 1300 miles in diameter.

Like that of Venus, the Martian atmosphere is almost entirely composed of carbon dioxide, with traces of nitrogen, argon and oxygen also present. The atmosphere is divided into three layers, which are (from the densest and closest): the *troposphere* which rises to 22 miles altitude, the *stratosphere* (22–80 miles) and *thermosphere* (80–140 miles) and the *exosphere* which accounts for the residual Martian atmospheric gases that exist above 140 miles altitude. Unlike the Earth's atmosphere, which generally circulates laterally with little interaction between the weather patterns of the northern and southern hemispheres, the Martian atmosphere has distinct north-south weather patterns that cross the equator. Circulating in what is called a Hadley cell (after its discoverer), warm Martian air rises in the hemisphere that is experiencing summer and moves to the opposite hemisphere at high altitude, where it sinks and then returns to the original hemisphere at low altitude. The Earth has Hadley cells too, but they don't cross the equator.

Mars has much less cloud cover than the Earth's roughly 50 percent coverage, but the cloud types are similar. These include cirrus and gravity wave clouds as well as cyclonic storms that on Earth would be referred to as hurricanes or typhoons. Low lying areas such as deep valleys or canyons develop ground fog as a result of frost being vaporized by the early morning Sun, a phenomenon not uncommon on the Earth.

While most of the Martian cloud cover is composed of water vapor (like the clouds of the Earth's atmosphere), carbon dioxide clouds exist at high altitudes and in the polar regions during the winter. Such clouds may result in precipitation in the form of dry ice snowstorms that play a role in replenishing the polar ice caps. The ice caps are also replenished by carbon dioxide condensation. While the carbon dioxide that is present in the ice caps as dry ice vaporizes and circulates within the Martian atmosphere, polar temperatures are such that most of the water ice present in the Martian ice caps remains frozen continuously even in summer.

Winds that are raised as a result of atmospheric circulation and seasonal or daily temperature fluctuation can result in huge dust storms in the Martian atmosphere. In 1977, for example, 35 major dust storms were recorded, and two of these developed into global storms.

Facing page: Olympus Mons rises 16 miles above the Martian surface—its summit caldera, shown here, is comprised of a series of overlapping craters, evidencing repeated eruptions over the many, many years.

Above left: This late winter picture, taken from the Viking 2 lander, shows traces of the white condensate which covered most of the planet's surface during both of the lander's winters on Mars. This condensate is composed of either water ice or carbon dioxide ice, or some combination of the two.

Above: Mars—the even now quite mysterious red planet—as viewed from Earth, with polar ice cap visible.

Successful Earth Expeditions to Mars

Name	Country of Origin	Launch	Date of closest contact or landing	Distance of closest contact (miles)
Mariner 4	USA	28 Nov 64	14 Jul 65	6103
Mariner 6	USA	24 Feb 69	31 Jul 69	2030
Mariner 7	USA	27 Mar 69	5 Aug 69	2180
Mars 2	USSR	19 May 71	27 Nov 71	Crashed
Mariner 9	USA	30 May 71	18 Nov 71	797.9
Mars 4	USSR	21 Jul 73	10 Feb 74	1364
Viking 1	USA	20 Aug 75	20 Jul 76	Landed
Viking 2	USA	9 Sep 75	3 Sep 76	Landed

Arcadia, Vastitas, Amazonis, and Acidalia

Below: Gigantic Olympus Mons is found in the Amazonis Planitia. Rising 15 miles above the surrounding plain, this volcano is the tallest mountain in the Solar System. The shaded area of the diagram *at right* indicates the portion of Mars which is mapped *on the facing page*—as it exists in relationship to the global map of Mars. Volcanism, faulting and extensive flooding have contributed to Mars' complex geologic makeup.

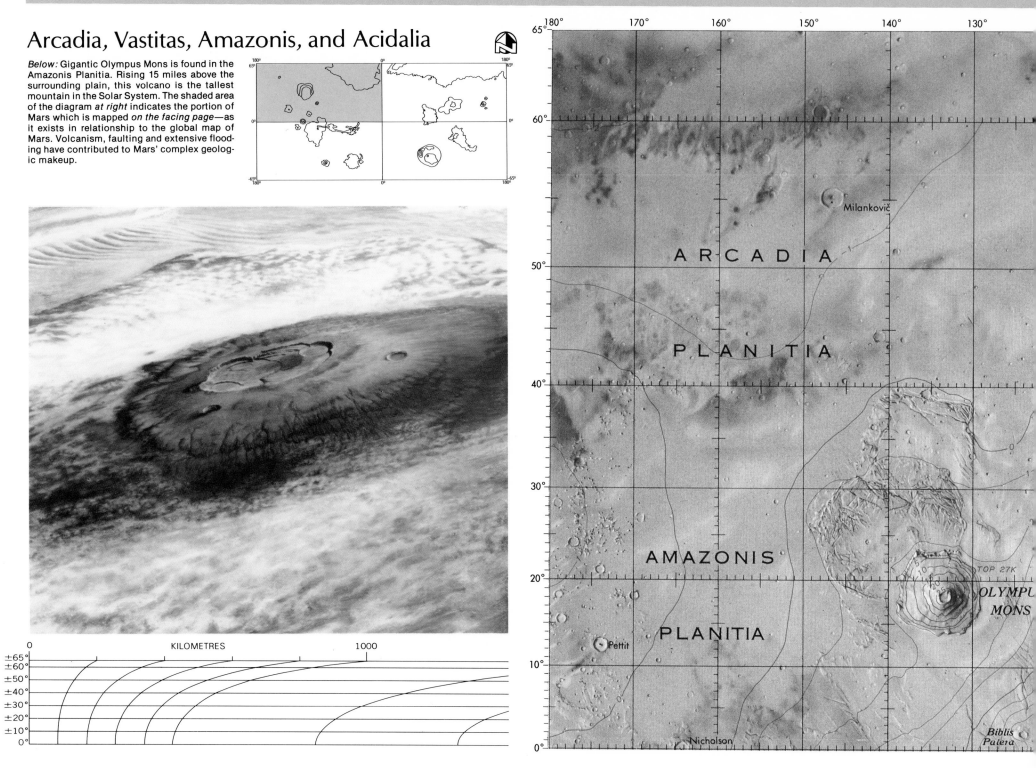

KILOMETRES

1000

±65°
±60°
±50°
±40°
±30°
±20°
±10°
0°

A R C A D I A

P L A N I T I A

AMAZONIS

PLANITIA

OLYMPUS MONS

Milankovič

Pettit

Nicholson

Biblis Patera

TOP 27K

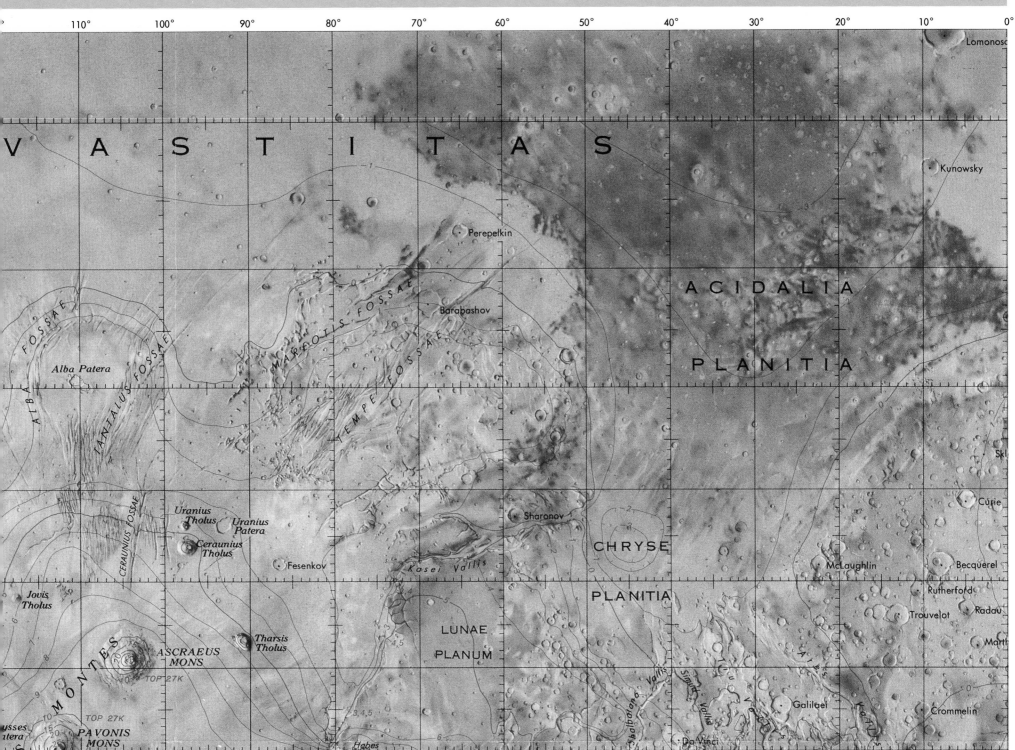

110°   100°   90°   80°   70°   60°   50°   40°   30°   20°   10°   0°

Lomonoso

V A S T I T A S

Kunowsky

Perepelkin

A C I D A L I A

FOSSAE

Barabashov

MAREOTIS FOSSAE

Alba Patera

PLANITIA

TANTALUS FOSSAE

TEMPE FOSSAE

ALBA

Sk

CERAUNIUS FOSSAE

Curie

Uranius
Tholus   Uranius
           Patera

Sharonov

CHRYSE

Ceraunius
Tholus

Becquerel

Fesenkov

Kasei Vallis

McLaughlin

Jovis
Tholus

PLANITIA

Rutherford

Radau

Trouvelot

Tharsis
Tholus

LUNAE

Mart

ASCRAEUS
MONS

TOP 27K

PLANUM

MONTES

TOP 27K

Galilaei

Crommelin

usses
tera

PAVONIS
MONS

Hebes

Da Vinci

Valles Marineris, Margaritifer and Argyre

Below: A Viking orbiter photo of Mars' huge seismic rift canyon, the 3000+ mile-long Valles Marineris, with (upper branch) Tithonium Chasma and (lower branch) Ius Chasma and above center of these, Ophir Chasma. The map *on the facing page* and those of the previous and following pages were created from Mariner 9 photos.

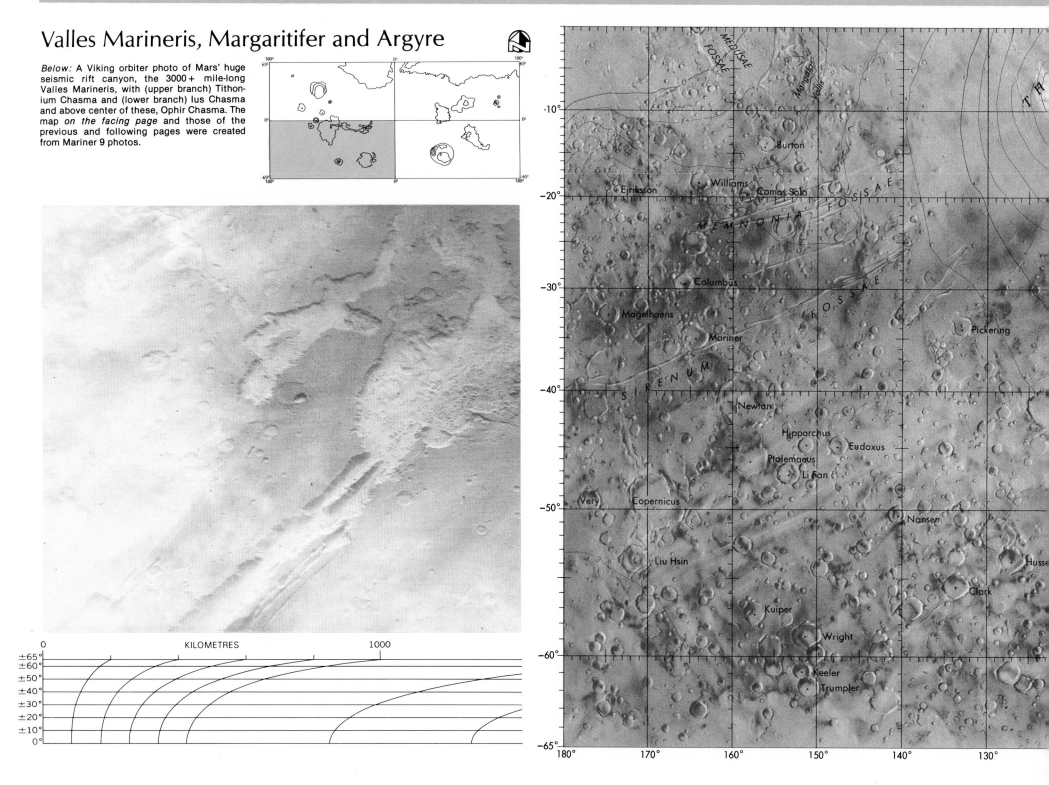

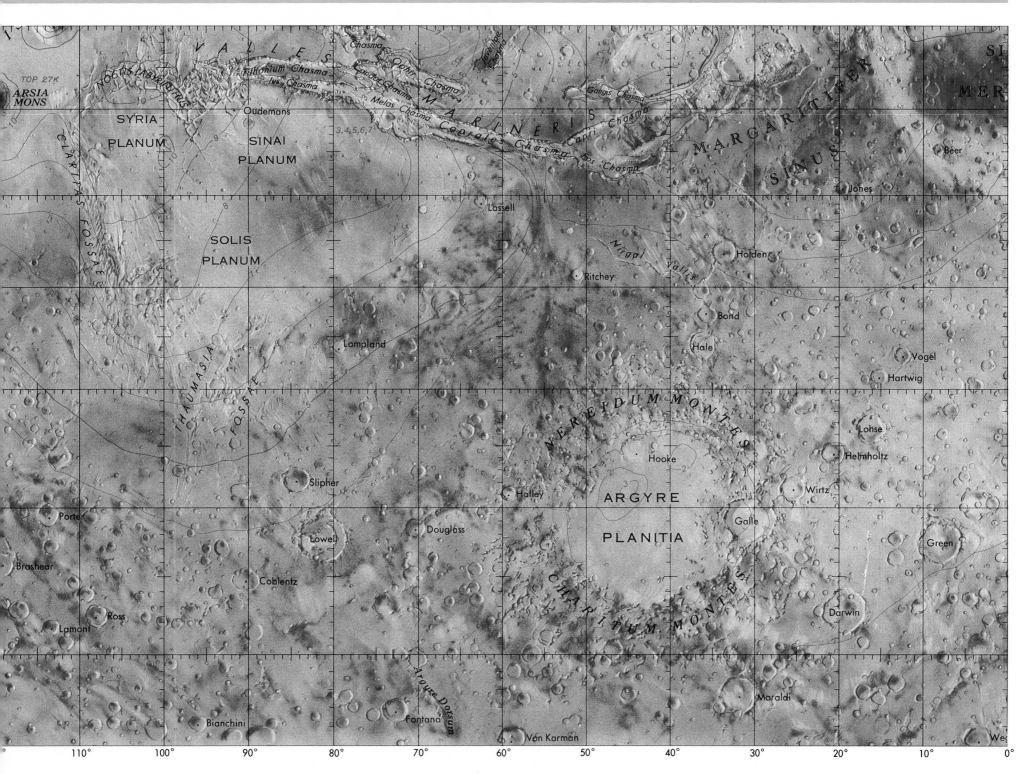

ARSIA
MONS

TOP 27K

VALLES

NOCTIS LABYRINTHUS

Tithonium Chasma

Ius Chasma

SYRIA
PLANUM

SINAI
PLANUM

CLARITAS FOSSAE

SOLIS
PLANUM

THAUMASIA FOSSAE

Oudemans

3,4,5,6,7

Chasma

Ophir Chasma

Candor Chasma

Melas Chasma

Coprates Chasma

Eos Chasma

Juventae Chasma

Gangis Chasma

Capri Chasma

MARINERIS

MARGARITIFER

SINUS

SI

MER

Beer

Jones

Lassell

Nirgal Vallis

Holden

Ritchey

Bond

Hale

Vogel

Hartwig

NEREIDUM MONTES

Lohse

Hooke

Helmholtz

Halley

ARGYRE

Wirtz

Slipher

Porter

Lowell

Douglass

PLANITIA

Galle

Green

Brashear

Coblentz

CHARITUM MONTES

Darwin

Ross

Lamont

Maraldi

Bianchini

Argyre Dorsum

Fontana

Von Karman

Weg

110°   100°   90°   80°   70°   60°   50°   40°   30°   20°   10°   0°

Borealis, Utopia and Elysium

Below: A summer afternoon on the 'Utopian Plain' photographed by the Viking 2 orbiter on 9 September 1976. Such plains as the Utopia Planitia were once covered by enormous lava floods, and were further refined by great water floods in Mars' enigmatic past.

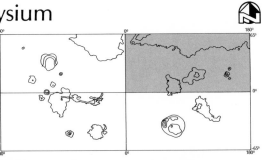

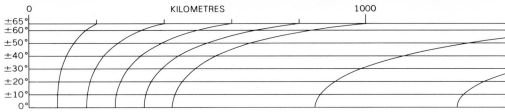

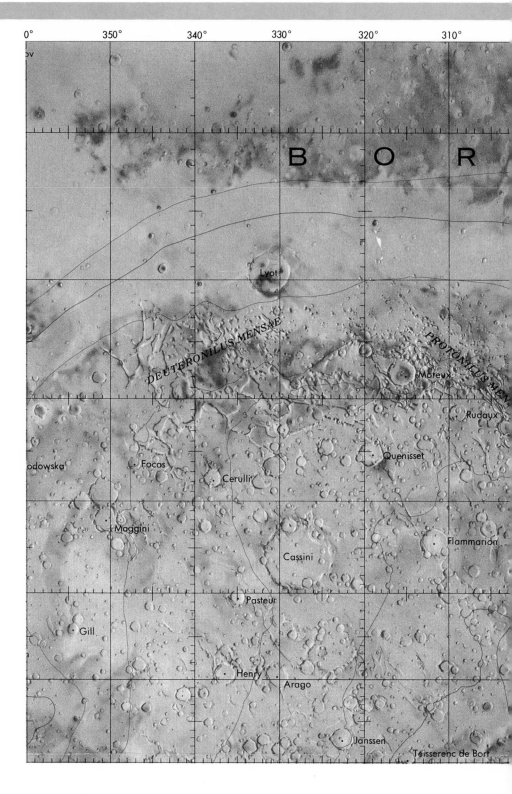

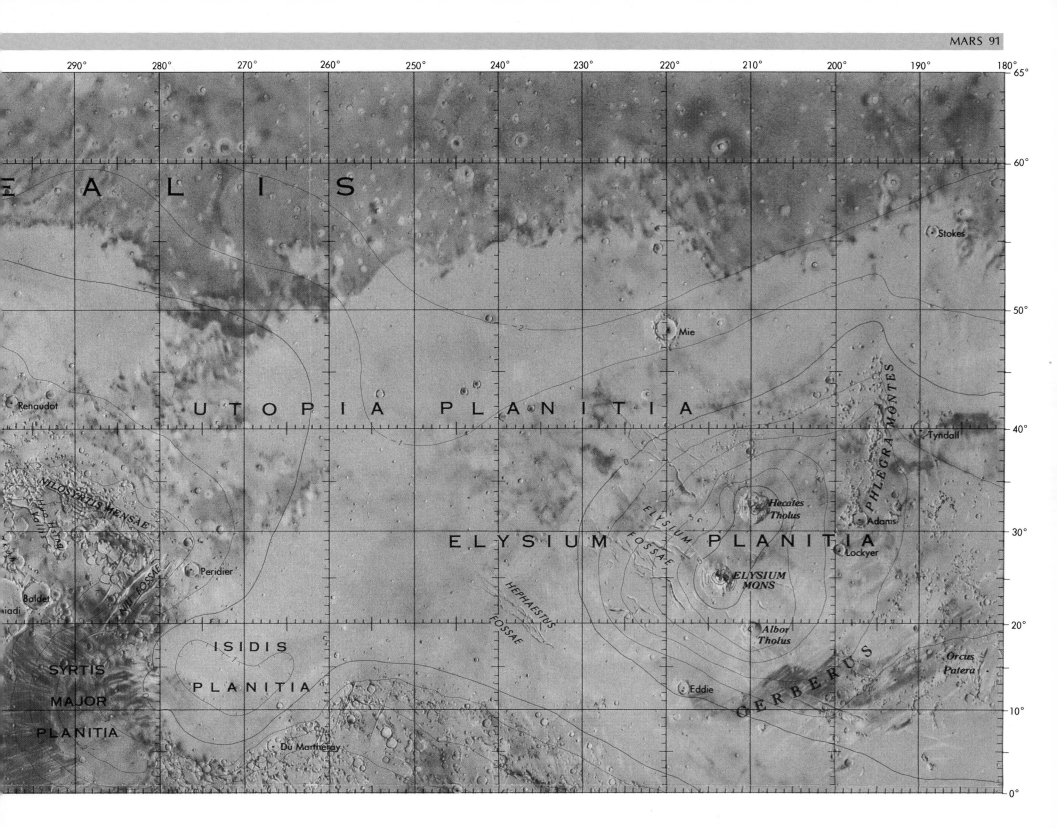

290° 280° 270° 260° 250° 240° 230° 220° 210° 200° 190° 180° 65°

60°

50°

40°

30°

20°

10°

0°

FI A L I S

Stokes

U T O P I A   P L A N I T I A

Renaudot

Mie

NILOSYRTIS MENSAE

Hsiang
Vallis

PHLEGRA MONTES

Tyndall

Peridier

ELYSIUM FOSSAE

Hecates
Tholus

Adams

E L Y S I U M   P L A N I T I A

Lockyer

NILI FOSSAE

Baldet

HEPHAESTUS

ELYSIUM
MONS

FOSSAE

ISIDIS

Albor
Tholus

Orcus
Patera

SYRTIS

PLANITIA

CERBERUS

MAJOR

Eddie

PLANITIA

Du Martheray

Hellas and Southeastern Quadrant

Mars' thin carbon dioxide atmosphere is plainly visible above the horizon in the spectacular photo *below.* In this view, we are looking out over the Argyre Planitia (the large basin in the foreground) in the direction of the Hellas Planitia. Some astro-geologists have theorized that such basins were created by asteroid impacts.

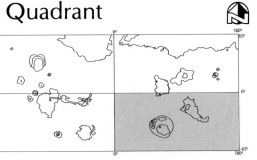

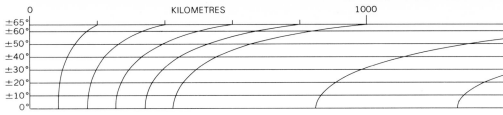

KILOMETRES

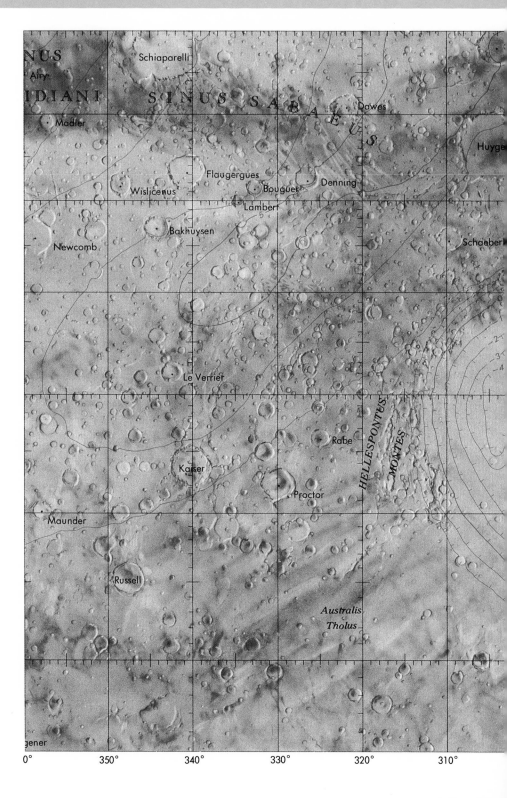

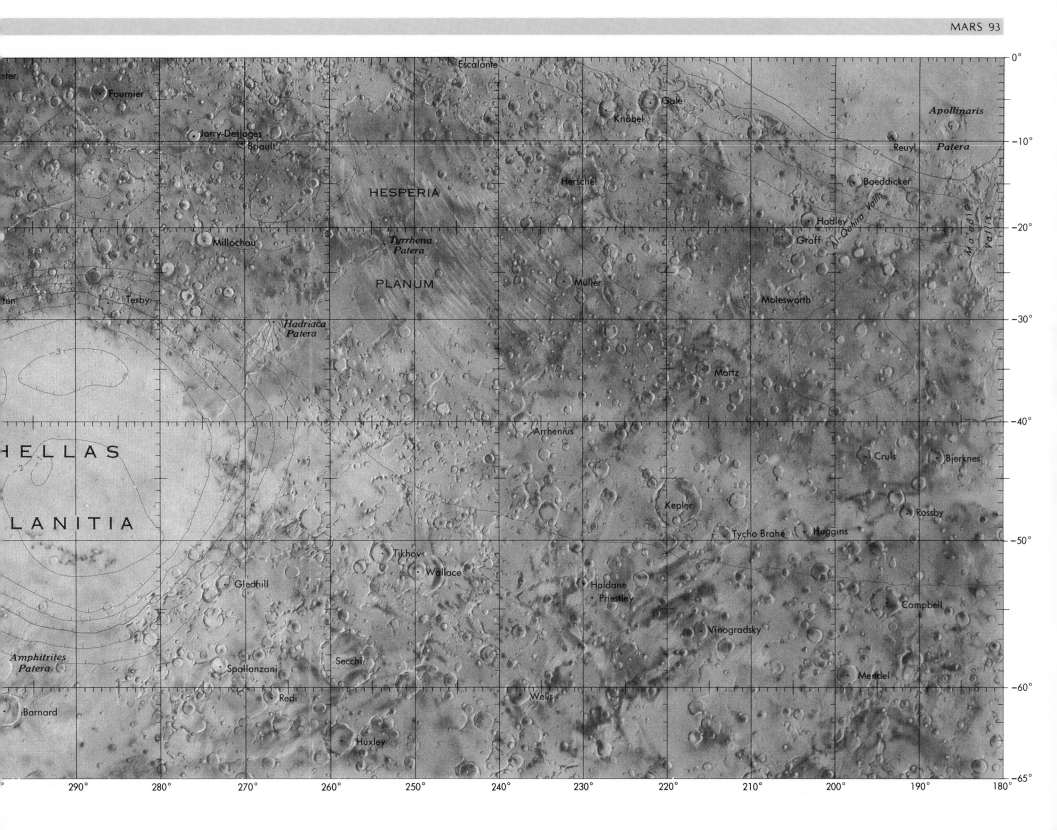

eter

Fournier

Jarry-Desloges
Briault

HESPERIA

Escalante

Gale

Knobel

Herschel

Apollinaris

Reuyl   Patera

Boeddicker

Hadley

Al-Qahira Vallis

Ma'adim Vallis

Millochau

Tyrrhena
Patera

Graff

PLANUM

Müller

Molesworth

Terby

Hadriaca
Patera

Martz

HELLAS

Arrhenius

Cruls   Bjerknes

LANITIA

Kepler

Rossby

Tycho Brahe   Huggins

Tikhov
Wallace

Haldane
Priestley

Campbell

Gledhill

Vinogradsky

Amphitrites
Patera

Secchi

Spallanzani

Mendel

Redi

Wells

Barnard

Huxley

290°   280°   270°   260°   250°   240°   230°   220°   210°   200°   190°   180°

0°

-10°

-20°

-30°

-40°

-50°

-60°

-65°

Geologic Mercator Projection

Below: This view along the Tharsis Ridge, which is the major volcanic region of Mars, was taken by the Viking 1 Orbiter on 13 July 1980. The volcanoes plainly seen here are, from bottom, Arsia Mons and Pavonis Mons, and at the top margin of this photo is Ascraeus Mons. At photo right is the western extent of Hoctis Labyrinthus, a region of dissecting faults.

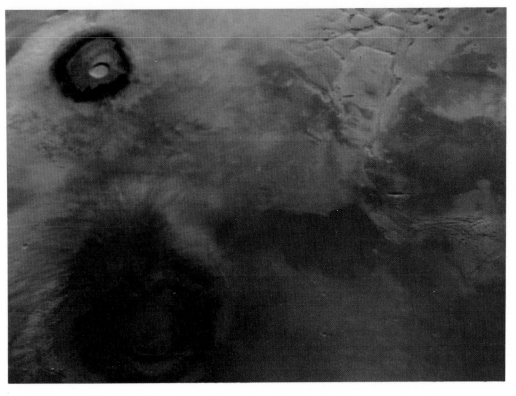

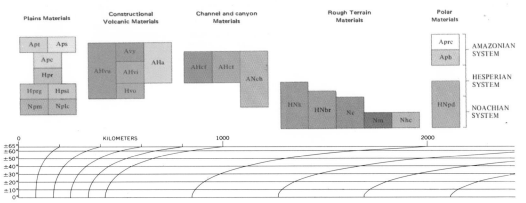

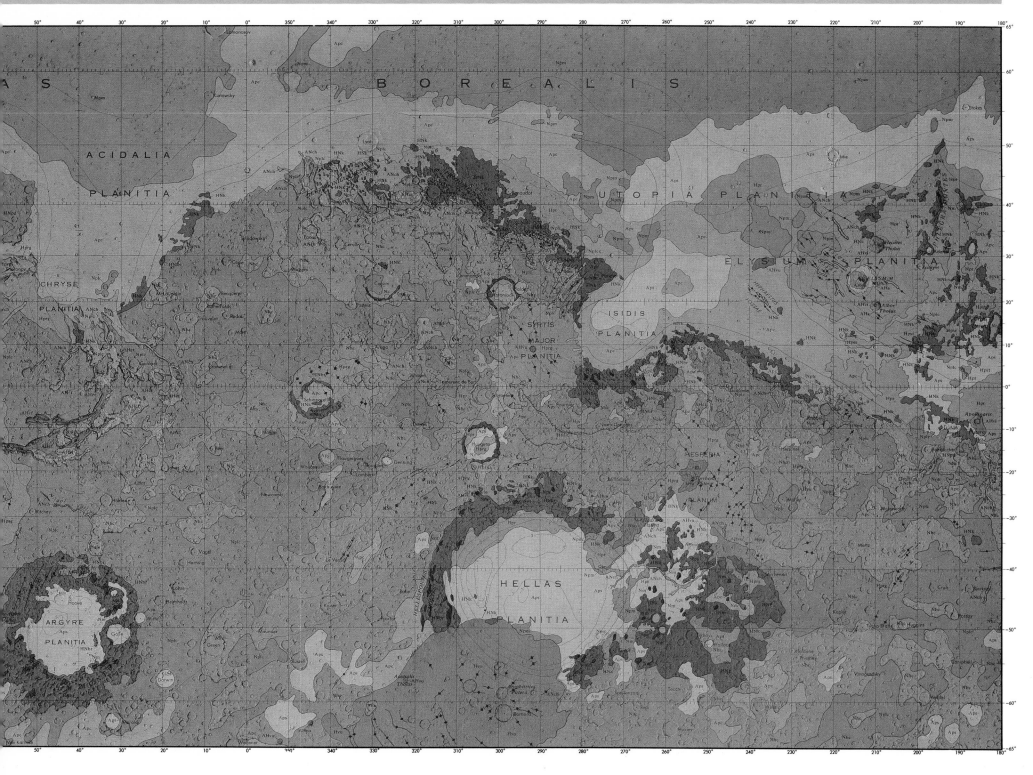

North Polar Topographical Map

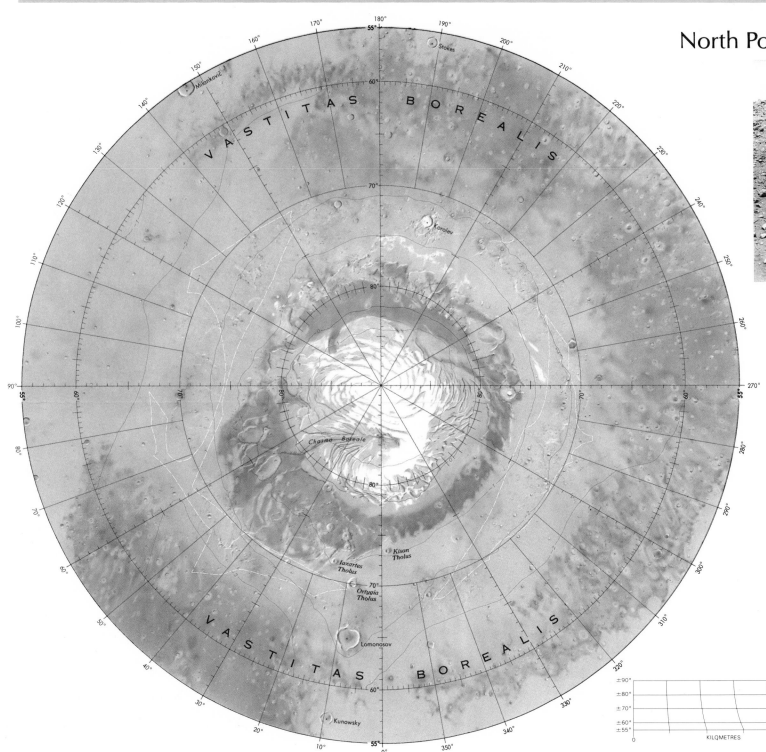

Please refer to the US Geologic Survey Mars maps on the previous pages to see how the USGS Mars polar region maps on these pages coincide with the 'global view' of Mars—all were based on Mariner 9 orbital photos taken on 4 August 1972. The main geologic periods on Mars are named for the global regions in which distinctive rock strata are found—such rock strata having uniquely formed as the result of various geomorphic processes which were dominant in given periods. Ergo, in descending order of age, we have the Noachian (rock strata) System—named for the Noachic quadrangle, in which the oldest rocks on the planet are the most exposed—the Hesperian System, named for the Hesperian Planum, located in the Mare Tyrrhenum quadrangle; and the Amazonian System, which includes, among other regions, the Martian ice caps, which are seen on *these* and the following *pages*.

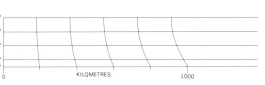

KILOMETRES

±90°
±80°
±70°
±60°
±55°

0   1000

North Polar Geologic Map

Above left is the first Viking lander photo of Mars, and *above* the same after color correction—ie, no more blue sky! The Martian sky's reddish cast is due to atmospheric scattering and reflection from wind blown dust particles in the lower atmosphere. The rocks visible here are probably limonite (hydrated ferric oxide). Widespread deposits of ferric (iron) oxide are responsible for the planet's overall reddish color.

Below: The North Polar Region's great ice cliffs. The North Polar ice cap is predominantly water ice while its southern counterpart is equal parts water and carbon dioxide ice.

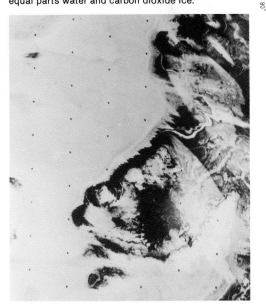

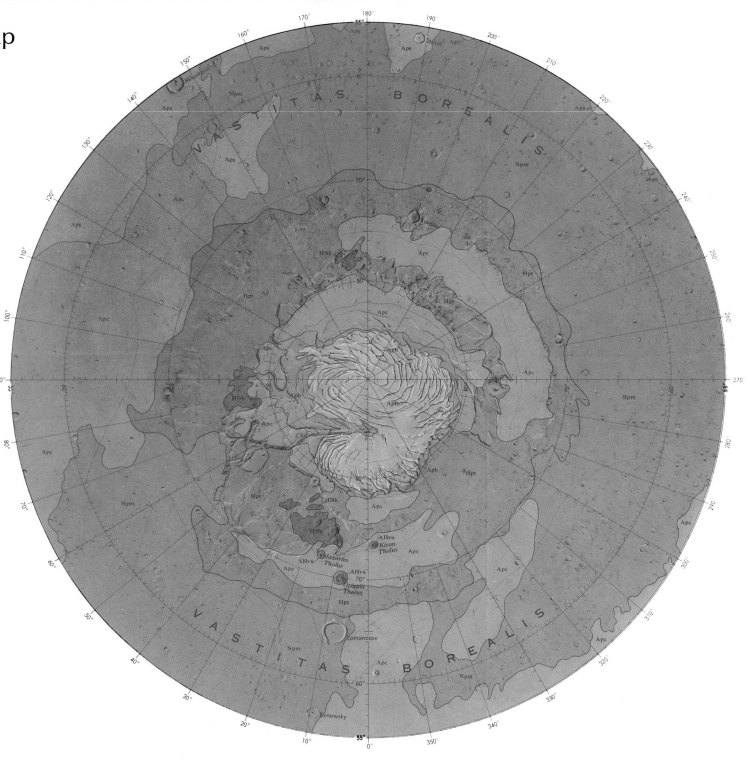

South Polar Topographical Map

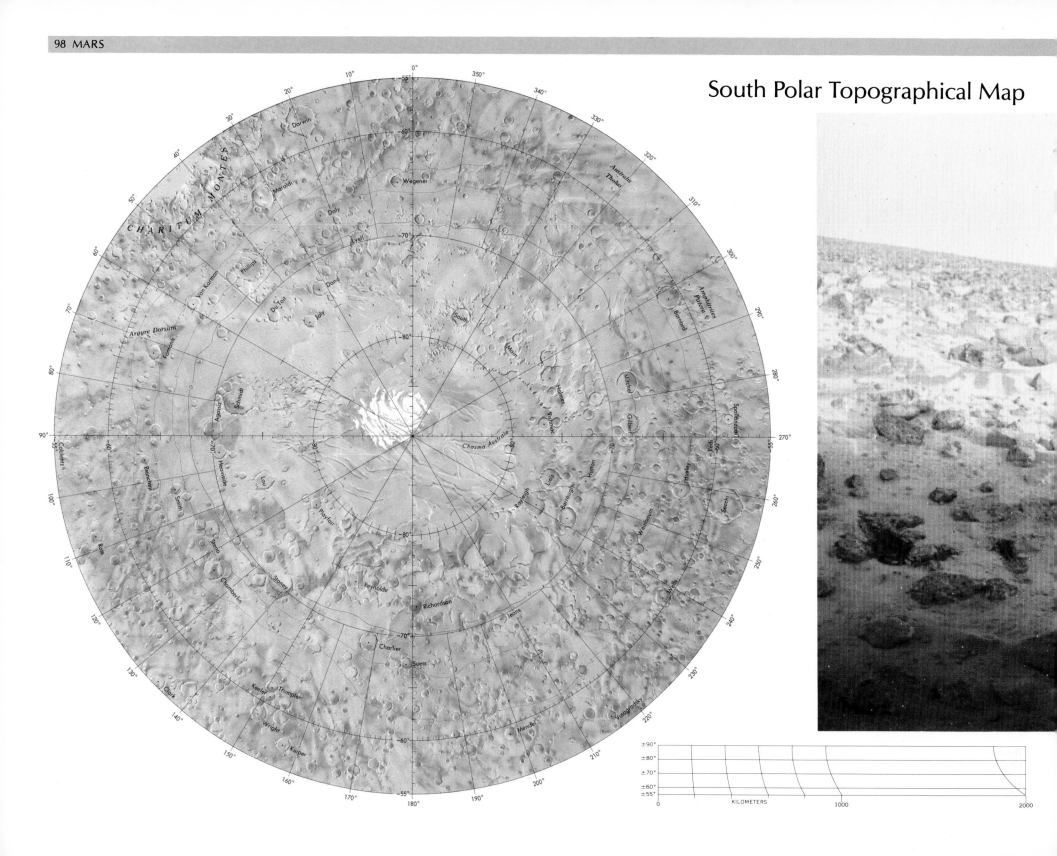

KILOMETERS

South Polar Geologic Map

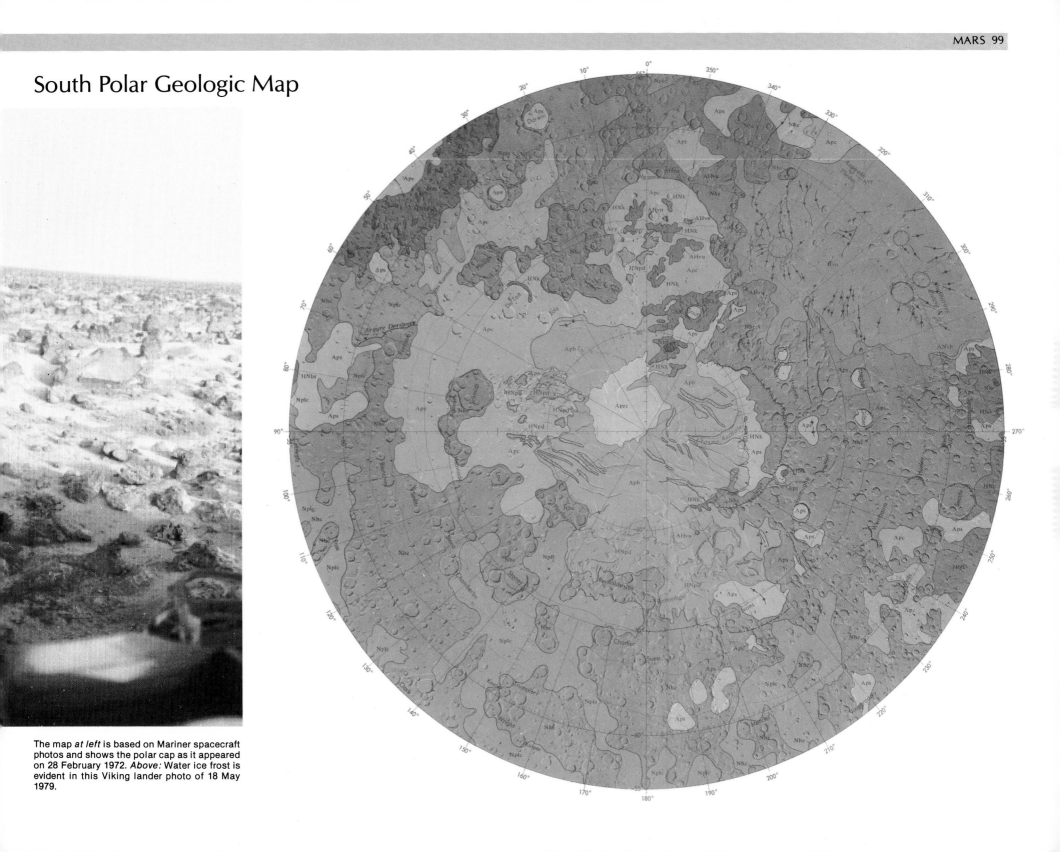

The map *at left* is based on Mariner spacecraft photos and shows the polar cap as it appeared on 28 February 1972. *Above:* Water ice frost is evident in this Viking lander photo of 18 May 1979.

LIFE ON MARS

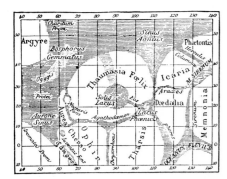

In 1659, the Dutch astronomer Christiaan Huygens (1629–1695) who was the first to identify a Martian surface feature (Syrtis Major) also calculated the Martian day to be almost the same as the Earth's, which it is. Seven years later the Italian astronomer Giovanni Domenico Cassini (1625–1712) discovered the Martian ice caps. By 1783 William Herschel (1738–1822) had correctly calculated the exact length of the Martian day and the exact inclination of Mars to its axis.

By the middle of the nineteenth century, the picture painted of Mars was that of a hospitable place that 'certainly' supported life in some form, probably similar to that of Earth. After all, their days were the same length and their seasons were parallel to ours. The darker areas on Mars were thought to represent vegetation and some observers recorded that this 'vegetation' waxed and waned with the Martian seasons.

In 1877, the Italian astronomer Giovanni Schiaparelli (1835–1910) made a startling discovery. There were channels, or *canali,* on the surface of Mars! Translated into English as 'canals,' the features were quickly ascribed to artificial origin. It was thought that intelligent creatures had constructed an intricate system of irrigation canals on Mars to bring water from the polar ice caps to the warmer equatorial region.

In 1894 the American astronomer Percival Lowell (1855–1916) opened his observatory at Flagstaff, Arizona primarily for the purpose of studying Mars. Lowell carefully observed and mapped the Martian surface and became a leading exponent of the idea that the canals were constructed by living creatures to irrigate their crops.

In the 1930s, Eugenios Antoniadi, a Greek-born astronomer working in France, produced a map of Mars which was quite accurate for its day, but one which rejected the earlier notion of artificial canals. By the late twentieth century the canal theory had been thoroughly discredited as having been an optical illusion, but the idea of Martian vegetation survived until Earth spacecraft visited the red planet.

The first spacecraft to pass near Mars was the American Mariner 4 in 1965 and it was followed by Mariner 6 and Mariner 7, four years later. The data returned by these flybys seemed to confirm the notion that Mars was a dull and lifeless place, roughly cratered and more like Mercury than the Earth.

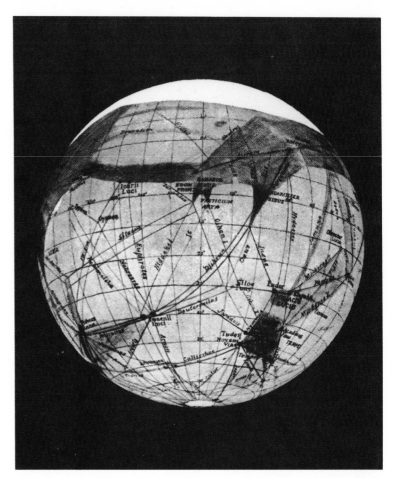

In 1971, however, the Mariner 9 spacecraft was placed into orbit around Mars. For the first time the full range of the planet's wonders, such as the great shield volcanoes and the vast networks of river beds, was revealed. Mariner 9 remained in service until October 1972, by which time the entire Martian surface had been mapped.

In August and September 1975, the United States launched the two identical Viking spacecraft toward Mars. Each Viking consisted of an orbiting module and a landing module designed to make a soft landing on the Martian surface. The Viking project was an outstanding success. The Viking 1 lander alighted in the Chryse Planitia on 20 July 1977 and continued to transmit data until November 1982. The Viking 1 orbiter conducted a close-up reconnaissance of the Martian moon Phobos and continued in its orbit around Mars until August 1980. The Viking 2 lander

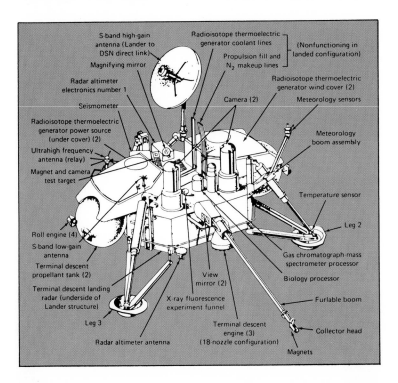

touched down in the Utopia region on 3 September 1976 and continued to transmit data until April 1980. The Viking 2 orbiter surveyed the Martian moon Deimos and continued to operate until it was powered down by Earth-based technicians in July 1978.

The Viking project expanded our knowledge of Mars manyfold and returned spectacular close-up photographs of the Martian surface that spanned Mars' seasonal changes for more than a Martian year. Viking answered a great many questions about Mars, but the notion of Martian life remains an enigma. There were three biology experiments aboard the Viking landers that were specifically designed to detect evidence of Martian life, but the answer returned was a resounding 'maybe not.' In each experiment, samples of Martian soil were scooped up by the landers' remote surface sample arms and brought aboard the spacecraft.

The Pyrolytic Release Experiment was designed to determine whether Martian organisms would be able to assimilate and reduce carbon monoxide or carbon dioxide as plants on Earth do. The easily monitored isotope carbon-14 was used and the results were described as 'weakly positive.' While the experiment could not be repeated by Viking on Mars, parallel experiments on

Earth showed that the same results could *possibly* be explained by chemical rather than biological reactions.

In the Labelled Release Experiment, an organic nutrient 'broth' was prepared and 'fed' to some samples of Martian soil, again using carbon-14 as the trace element. If micro-organisms were present, they would 'breathe out' carbon dioxide as they 'ate' the nutrients. Carbon dioxide was, in fact, detected! However the outgassing of carbon dioxide stopped and could not be restarted. This could have indicated some sort of chemical reaction or that a microbe *had* been present but that it had died while 'eating' the 'broth.' To distinguish a chemical reaction from a biological reaction, the mixture was heated. This process stopped whatever it was that was producing the carbon dioxide, which *should* have ruled against the notion of a chemical reaction, but which might confirm that it had been caused by a now-deceased organism. In the end, the Labelled Release Experiment was labelled inconclusive because the activities of whatever produced the carbon dioxide had no exact parallel with known reactions of Earth life.

The Gas Exchange Experiment was designed to examine Martian soil samples for evidence of gaseous metabolic changes, by again mixing a sample with a nutrient 'broth.' Because the Martian environment is so dry, it was decided to gradually humidify the samples before plunging them into the 'broth' so as not to 'shock' any of the life forms that might be present. A major shock came instead to the Earth-based experimenters as the sample was being humidified—there was a sudden burst of oxygen! When the nutrient 'broth' itself was added, there was some evidence of carbon dioxide but no more oxygen. Once again the results were described as 'inconclusive' because the results could not be explained by known biological reactions. Subsequent studies have been done to attempt to determine whether some type of oxydizing agent exists in the Martian soil which could provide a 'chemical reaction' explanation for the strange results of the Gas Exchange Experiment.

The three Viking biology experiments raised some curious questions, but there remain no conclusive answers to the question that has intrigued Earthbound observers. Perhaps the answer lies closer to the Martian poles where there is more water, or perhaps the question might be restated as whether life *might have existed* at one time on Mars. In the long-gone days when rivers ran on the Martian surface, did some civilization or even just a species of moss flourish on their banks? Will paleontologists or archeologists from Earth one day discover fossils or ruins amid the drifting rust-red Martian sands?

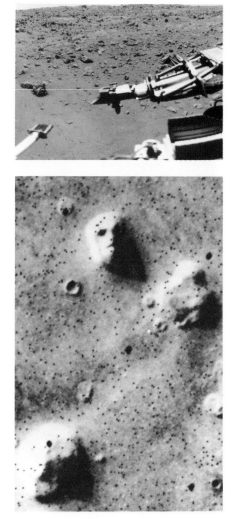

Above left: A diagram of the Viking landers. They did more to help explain Martian life than any other project, but the jury is still out.

At top, above: The Viking 1 soil sampler, shown here, gathered material with which to conduct tests for Martian life forms.

The eyes of Mars are upon you: the one mile wide rock formation *above* strongly suggests a human face, but is, we're told, nothing but a shadow trick.

THE MARTIAN MOONS

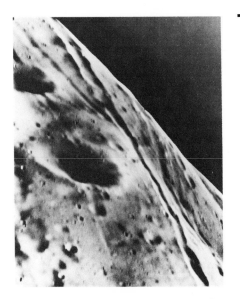

Viking Orbiter 1 took the photo *above* of a region in the northern hemisphere of Phobos that has striations and is heavily cratered. The striations, which appear to be grooves rather than crater chains, are about 328 to 656 feet wide and dozens of miles long. Craters range in size from 32 feet to ¾ mile in diameter.

A pair of images of Deimos from the Viking 1 orbiter—one taken through the camera's violet filter, the other through the orange filter—were combined in the single image *at right*. Deimos is actually a uniform gray color.

These Viking 1 orbiter pictures (shown sequentially, left to right, at *below right*) were part of an experiment to locate the position of the Viking 1 lander on Mars using shadows of the moons Phobos and Deimos. At the same time, the Viking 1 lander took pictures as the shadow of Phobos crossed over it. Careful timing and detailed processing of these and similar Phobos/Deimos shadow pictures will allow scientists to locate the lander within 0.6 mile. The crater Sharonov, 93 miles in diameter is at top, Chryse Planitia, where the Viking 1 lander touched down on 20 July 1976, is at the right.

In 1877, as Schiaparelli was astounding the world with his discovery of *canali* on the Martian surface, the American astronomer Asaph Hall (1829-1907) made an even more important discovery. Observing Mars from the US Naval Observatory in Washington, Hall determined that the red planet was accompanied by two tiny moons. Named Deimos (terror) and Phobos (fear) after the charioteers of (some sources say 'the horses of') the mythological war-god *Mars,* the two moons are irregularly-shaped rocks pocked with numerous craters. Phobos, the larger of the pair, is just over 17 miles in length, while Deimos is less than nine miles long. Because of their shape, size and texture, it is thought that they originated among the asteroids and became trapped in Martian orbit at the time of the formation of the Solar System. Neither Deimos nor Phobos has the mass to allow them to hold an atmosphere, but they exert sufficient gravity to retain a thin layer of dust on their surfaces—which is perhaps residue from the meteorite impacts that caused the cratering.

Both are relatively small, and Deimos would appear no larger from the Martian surface than Venus does from the Earth. Phobos, however, would appear as if it were one third the size of the Earth's Moon. Though both moons revolve around Mars in the same direction, Phobos would appear to revolve in the *opposite* direction because it revolves in less than eight hours, a third of the time that it takes Mars to rotate on its axis.

Phobos is characterized by a number of large craters, the largest of which, Stickney (the maiden name of Asaph Hall's wife), is six miles across. The next largest craters are Hall and Roche

Data for Deimos

Diameter: 6.2 × 7.5 × 9.9 miles (10 × 12 × 16 km)
Distance from Mars: 14,540 miles (23,400 km)
Mass: $9.10 × 10^{15}$ lb ($20 × 10^{15}$ kg)
Rotational period (Deimosian day): 30.35 Earth hours
(1 day, 6 hrs, 21 min, 16 sec)
Sidereal period (Deimosian year): 30.35 Earth hours
(1 day, 6 hrs, 21 min, 16 sec)
Inclination to Martian equatorial plane: 1.8°

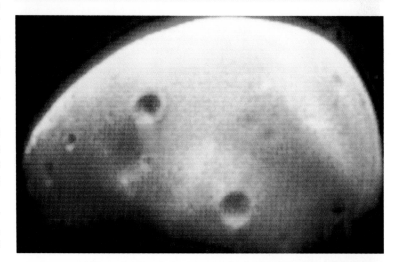

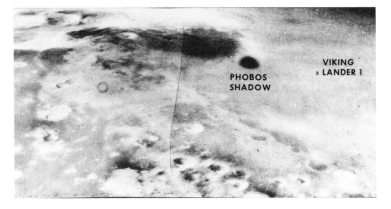

PHOBOS
SHADOW

VIKING
x LANDER 1

THIN CLOUD LAYER

VIKING
LANDER 1

PHOBOS
SHADOW

which are about half the size of Stickney, while the average Phobos crater diameter is about 500 feet. Other features include surface fractures that were probably induced by Martian gravitational effects.

Deimos has a much smoother surface than its larger brother, with fewer craters, and none with diameters greater than two miles. In other words, the ratio of longest overall dimension to largest crater diameter is 2.8 to 1 on Phobos and 5.3 to 1 on Deimos. Because Deimos is more than twice as far from the parent planet than is Phobos, it shows no evidence of the surface fracturing that has been induced on the latter by the tidal effects of Martian gravity.

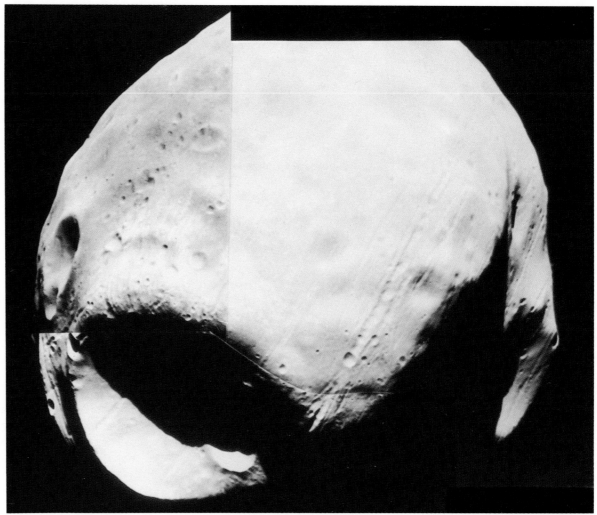

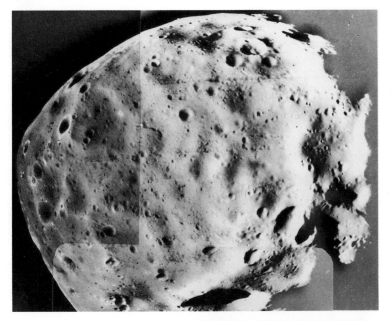

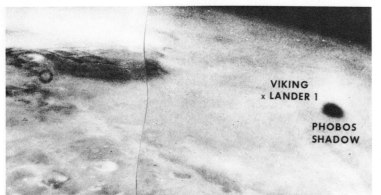

Data for Phobos

Diameter: 12.4 × 14.3 × 17.4 miles (20 × 23 × 28 km)
Distance from Mars: 5760.11 miles (9270 km)
Mass: 4.36 × 10¹⁵ lb (9.6 × 10¹⁵ kg)
Rotational period (Phobian day): 7.66 Earth hours
Sidereal period (Phobian year): 7.66 Earth hours
(7 hrs, 39 min, 27 sec)
Inclination to Martian equatorial plane: 1.1°
Largest surface feature: Stickney (a large crater)
6 miles in diameter

Above left: The Viking 1 orbiter flew within 300 miles of Mars' inner satellite, Phobos, to obtain this mosaic of the asteroid-sized moon on 18 February 1977.

Above: This photomosaic shows the side of Phobos which always faces Mars. North is at the top. Stickney, the largest crater on Phobos, is at the left near the morning terminator. Kepler Ridge is casting a shadow in the southern hemisphere which partly covers the large crater, Hall, at the bottom.

JUPITER

Justifiably named for the king of all the Roman gods, Jupiter is the largest planet in the Solar System, and second in mass only to the Sun. This 'king' of planets has 1330 times the volume of the Earth and 318 times the mass. As the Solar System was being formed 4.5 billion years ago, Jupiter may have had the makings of becoming a star. At that time it was 10 times its present diameter and heated by gravitational contraction. It may have blazed like a second Sun. Had the nuclear reactions within Jupiter become self sustaining as they did in the Sun, the two objects may have become a double star of the type that exists elsewhere in the galaxy—and the Solar System would have been a vastly different place than it is today. But Jupiter failed as a star and gradually began to cool, and to collapse to its present size. As Jupiter cooled it became less brilliant, so that after a million years the 'star that might have been' went from one hundred thousandth the luminosity of the Sun to one ten billionth. Today, however, as much energy is still radiated from *within Jupiter itself* as it *receives from the Sun.*

Jupiter's magnetic field is more than 4000 times greater than that of the Earth, leading observers to postulate the existence of a metallic *core.* Strangely, the center of Jupiter's magnetic field is not located at the planet's center, but at a point 6200 miles offset from the center and tipped by 11 degrees from the rotational axis. This point is also the center of Jupiter's vast *magnetosphere,* which is six million miles across. Three million miles from Jupiter the plasma reaches the hottest temperatures record-

ed in the Solar System, roughly 17 times as hot as the interior of the Sun.

Like the Sun, Jupiter is composed almost entirely of hydrogen and helium, and unlike the terrestrial planets, it may be composed almost entirely of gases and fluids with no solid surface. If there is a solid rocky surface, the solid diameter of Jupiter may actually be just slightly larger than the Earth. Above this rocky surface, if it exists, there may be a layer of ice more than 4000 miles thick which is kept frozen by *pressure* rather than temperature, as the *temperatures* would be frightfully hot.

There is almost certainly a sea of liquid metallic hydrogen that makes up the bulk of Jupiter. Above the liquid metallic hydrogen is a transition zone leading to a layer of fluid molecular hydrogen. Jupiter's *atmosphere* is characterized by colorful swirling clouds that cover the planet completely. These clouds form in Jupiter's *troposphere,* at the altitude where convection takes place. The lower clouds, like those of the Earth, are thought to be composed of water vapor with ice crystals being present at higher altitudes. Above these, higher clouds are composed of ammonium hydrosulfide, with Jupiter's high 'cirrus clouds' composed of ammonia. At a point 40 miles above the ammonia cirrus, where the Jovian troposphere gives way to the *stratosphere,* temperatures can dip to colder than −150 degrees Fahrenheit. Above that, in the *ionosphere,* however, temperatures increase again.

Jupiter's atmosphere is a complex and dynamic feature char-

Facing page: The king of planets is shown here as it appeared to the Voyager interplanetary probe. The gaseous Jupiter, 'the star that almost was,' is in fact the largest planet in the Solar System.

Above: Jupiter as seen from Earth. It can be viewed from good old *terra firma* with a fairly low-powered telescope.

acterized by distinct horizontal 'belts' or darker bands of clouds that exist at semi-symmetrical intervals in the northern and southern hemispheres and which alternate with lighter colored 'zones.'

The most outstanding feature on Jupiter is certainly the Great Red Spot. First observed in 1664 by the astronomer Robert Hooke (1653-1703), it is a brick red cloud three times the size of the Earth. Described as a high pressure system, the Great Red Spot resembles a storm and exists at a higher and colder altitude than most of Jupiter's cloud cover, although traces of ammonia cirrus are occasionally observed above it. It rotates in a counter-clockwise direction making a complete rotation every six Earth days, and it varies slightly in latitude. The exact nature of the Great Red Spot is uncertain, but one theory is that it is above an updraft in the Jovian atmosphere in which phosphine, a hydrogen-phosphorus compound, rises to high altitudes—where it is broken down into hydrogen and red phosphorus-4 by solar ultraviolet radiation. The pure phosphorus would give the Great Red Spot its characteristic color. Another theory has the Great Red Spot as the top of a column of stagnant air that exists above a topographical surface feature far below, within Jupiter. The Great Red Spot is almost certainly the top of some sort of high altitude updraft plume from below the Jovian cloud cover, but the divergent flow from it is quite small. For example, one smaller feature was seen to circle the Great Red Spot for an Earth month without altering its distance.

The Great Red Spot may be an awesome feature, but it is a transient one. Any storm that has been raging for more than 300 years can certainly be termed an impressive meteorological phenomenon, but it hasn't been constant in its intensity. Between 1878 and 1882, it was seen as very prominent, but thereafter it dimmed markedly until 1891. Since then, it has waned slightly several times—in 1928, 1938 and again in 1977.

Other intriguing meteorological phenomena have also been observed in the Jovian atmosphere, including smaller red spots in the northern hemisphere and some dark brown features that formed at the same latitude as the Great Red Spot. Designated as the South Tropical Disturbance, these features were first observed in 1900, overtook and 'leaped' past the Great Red Spot several times and gradually began to fade in 1935, disappearing five years later.

In 1939, a group of large white spots formed near the Great Red Spot in the southern hemisphere. Like their larger red counterpart, they rotate counterclockwise. Similar but smaller features have been observed in the northern hemisphere, where they

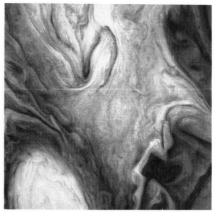

Facing page: The Great Red Spot is Jupiter's most striking optical feature. It is probably the top of an atmospheric plume composed of phosphine, which breaks down when exposed to outer-atmospheric solar radiation.

Looking very much like an Art Deco tabletop, the region near the Great Red Spot evidences the several 'white spots' that formed there in 1939. In the photo *at left* and *below left,* which were taken almost four months apart by Voyagers 1 and 2, note how these white spots migrate in relation to the Great Red Spot. Such migration adds strength to the assertion that the Jovian atmosphere actually moves at widely varing speeds, as its many currents or winds follow discrete latitudinal paths.

Above: A closeup of Jupiter's turbulent cloud cover.

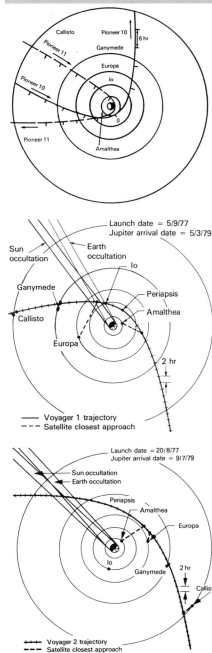

The routes to Jupiter: *From the top:* Pioneers 1 and 2; Voyager 1 (*photo this page*); and Voyager 2. *Opposite:* Jovian rings, backlit.

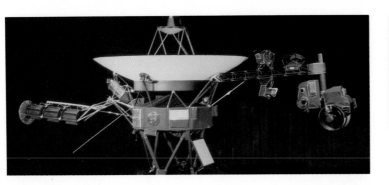

are seen to rotate in a clockwise direction. All of these features are oval in shape in the *tropical* and *temperate* latitudes, but are more rounded in polar regions. Like the Great Red Spot, the lesser white spots appear to be the tops of some sort of plumes, surrounded by darker filamentary rings.

Jupiter's equatorial band is characterized by regularly spaced features that resemble the type of convective storms that originate in the tropical latitudes on the Earth.

In the northern hemisphere, small brown counter-clockwise-rotating storms race and tumble through the tropical and temperate zones. These playful features may collide, combine and later break apart.

The most notable interactions between Jovian clouds are in the region of the Great Red Spot. Other spots, or storms, caught on its outer edge might break in two with one piece remaining in the vortex and the second moving away in the same direction as that of the original storm. Occasionally, a ribbon of white clouds might form around the periphery of the Great Red Spot.

In the north polar region there is a very prominent aurora resulting from ultra violet glows of atomic and molecular hydrogen. Nighttime observations by the Voyager spacecraft also observed widespread clusters of electrical storms at all latitudes in the Jovian atmosphere.

Unknown before the close-up observations by the Voyager spacecraft in 1979, Jupiter has a distinct *ring system*. Unlike the very visible Saturnian rings, the Jovian rings are very thin and narrow, and are not visible except when viewed from behind the night side of the planet, when they would be backlighted by the Sun. The ring system is divided into two parts that begin 29,000 miles above Jupiter's cloud tops, although some traces of ring material exist below that altitude. The two parts are a faint band 3100 miles across, feathering into a brighter band 500 miles across. The rings are composed of dark grains of sand and dust and are probably not more than a mile thick.

Data for Jupiter

Diameter: 88,650 miles (142,984 km)
Distance from Sun: 505,734,000 miles (815,700,000 km) at aphelion
459,358,000 miles (740,900,000 km) at perihelion
Mass: 8.632×10^{26} lb (1.899×10^{27} kg)
Rotational period (Jovian day): 9.84 Earth hours
Sidereal period (Jovian year): 4333 Earth days
Eccentricity: 0.048
Inclination of rotational axis: 3.12°
Inclination to ecliptic plane (Earth = 0): 1.3°
Albedo (100% reflection of light = 1): .34
Mean temperature: 26,637° F
Maximum temperature: 53,476° F
Minimum temperature: Great Red Spot: less than −202° F
Largest feature: The Great Red Spot 16,280 × 8575 miles
Major atmospheric components: Hydrogen (±90%)
Helium (±10%)
Other atmospheric components: Methane, Ammonia, Ethane, Acetylene, Water vapor, Phosphine, Carbon monoxide, Germanium, Tetrahydride

The Jovian Ring System

	Ring width (in miles)	Orbital Distance (Planet center to inner edge of ring in miles)	Ring Composition
Secondary	31,869.24	44,266.76	dust-like particles
Primary	3968	76,136	dust-like particles

Successful Earth Expeditions to Jupiter

Name	Country of Origin	Launch	Date of closest contact or landing	Distance of closest contact (miles)
Pioneer 10	USA	2 Mar 72	4 Dec 73	80,786
Pioneer 11	USA	5 Apr 73	2 Dec 74	26,536
Voyager 2	USA	20 Aug 77	10 Jul 79	398,660
Voyager 1	USA	11 Sep 77	5 Mar 77	172,360

To explain the fact that Jupiter gives off more energy than it receives from the Sun, pre-Voyager spaceflight theories postulated the Jupiter cross-sectional model illustrated *at right;* which is in contrast to the Jovian model illustrated *at far right,* which is based on the Voyager probe's discoveries.

Below: Ten three-color images, obtained during a single Jovian rotation on 1 February 1979, were used to form this mosaic 'Mercator projection' of Jupiter. The zones of the Jovian atmosphere are shown thus: NTeZ (North Temperate Zone), NTrZ (North Tropical Zone), NEB (North Equatorial Belt), EZ (Equatorial Zone), SEB (South Equatorial Belt), STrZ (South Tropical Zone) and STeZ (South Temperate Zone).

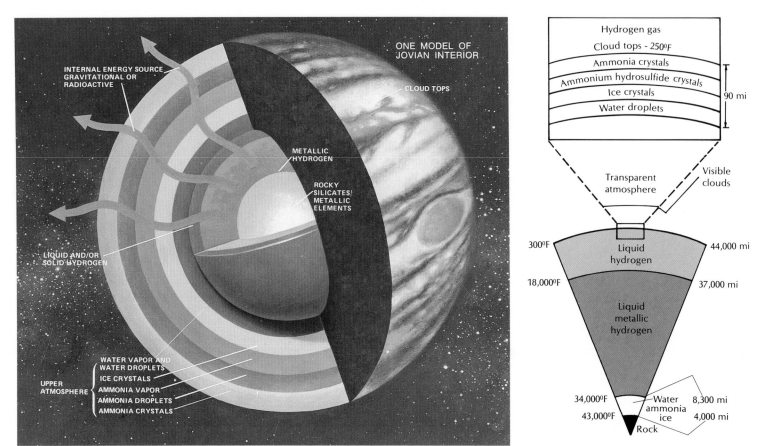

ONE MODEL OF JOVIAN INTERIOR

INTERNAL ENERGY SOURCE GRAVITATIONAL OR RADIOACTIVE

CLOUD TOPS

METALLIC HYDROGEN

ROCKY SILICATES/ METALLIC ELEMENTS

LIQUID AND/OR SOLID HYDROGEN

WATER VAPOR AND WATER DROPLETS
ICE CRYSTALS
AMMONIA VAPOR
AMMONIA DROPLETS
AMMONIA CRYSTALS

UPPER ATMOSPHERE

Hydrogen gas
Cloud tops - 250⁰F
Ammonia crystals
Ammonium hydrosulfide crystals
Ice crystals
Water droplets

90 mi

Transparent atmosphere

Visible clouds

300⁰F   Liquid hydrogen   44,000 mi

18,000⁰F   37,000 mi

Liquid metallic hydrogen

34,000⁰F   Water ammonia ice   8,300 mi

43,000⁰F   4,000 mi

Rock

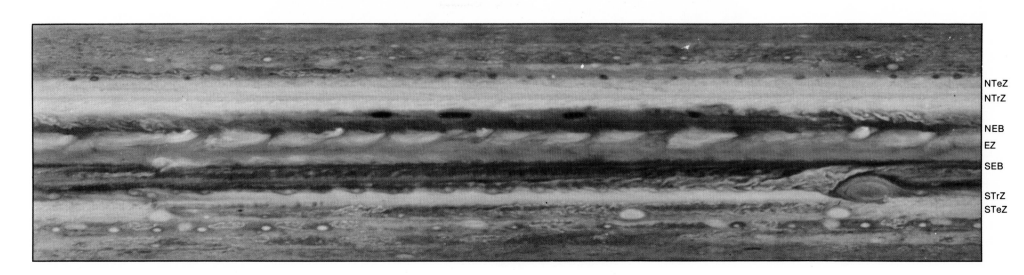

NTeZ
NTrZ

NEB

EZ

SEB

STrZ
STeZ

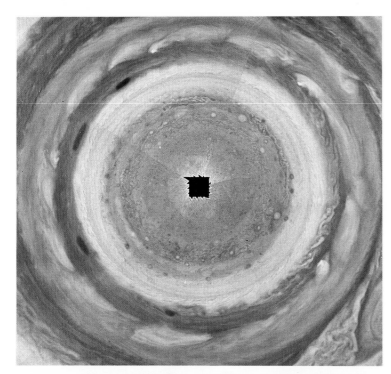

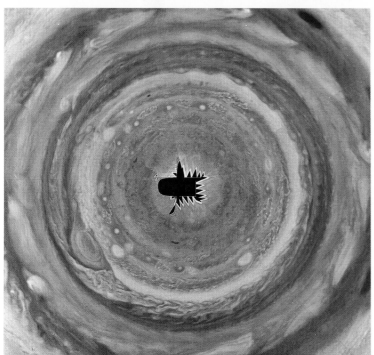

JUPITER'S MOONS

The 16 known Jovian moons are organized into a very orderly system of four dissimilar groups, each comprised of four similar sized moons orbiting in distinctly different planes.

1. The inner group (except for Amalthea) were all discovered during the Voyager project, and they have diameters of less than 200 miles. They all orbit in a plane whose orbital inclination is less than half a degree and they are located less than 140,000 miles from Jupiter.

2. The second group, called 'The Galileans,' were all discovered in 1610 by Galileo Galilei (1564–1642) and they have diameters greater than 1900 miles. They all orbit in a plane whose orbital inclination is less than half a degree and they are all between 250,000 and 700,000 miles from Jupiter.

3. The third group were all discovered in the twentieth century prior to Voyager, and they all have diameters of less than 105 miles. They all orbit in a plane whose orbital inclination is between 26 and 29 degrees and they are all between 6.9 and 7.2 million miles from Jupiter.

4. The final group were all discovered in the twentieth century prior to Voyager, and they all have diameters of less than 17 miles. They all orbit in a plane whose orbital inclination is between 147 and 163 degrees and they are all between 12.8 and 14.7 million miles from Jupiter.

The inner group of Jovian moons is dominated by Amalthea, discovered in 1892 by Edward Emerson Barnard (1857–1923) using the 36 inch refractor telescope at Lick Observatory. Amalthea is named for the goat-like nurse of Zeus (the Greek equivalent of the Roman god Jupiter). In Greek mythology Zeus broke off one of Amalthea's horns and endowed it with power to be filled with anything the owner wished, causing Amalthea to become associated with prosperity and riches. Elongated by gravitational pressure from Jupiter and heated by the Jovian magnetosphere, Amalthea has a reddish sulfurous surface like that of the moon Io. This surface is dominated by the craters Pan and Gaea and two mountains, Ida and Lyctas.

The innermost Jovian moon Metis was originally named 1979 J3 when it was first discovered in Voyager photographs. It is named for the Greek god of prudence, wife of Zeus and mother of Athena. The other two inner Jovian moons, Adrastea and Thebe, were originally designated 1979 J1 and 1979 J2 and were the first two new moons in the Solar System to be discovered by

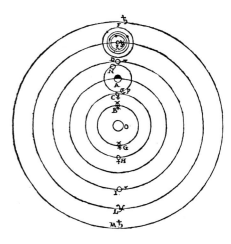

Above: Galileo's drawing of the Copernican view of the Solar System published in 1632 includes the Jovian satellites Io, Europa, Ganymede and Callisto, that he had discovered. The other features include A, Earth; O, Sun; B/G, Mercury; C/H, Venus; N, Full Moon; P, Quarter Moon; D/I, Mars; E/L, Jupiter with its four Galilean moons; and F/M, Saturn.

This page, above left: This photo was constructed from a series of consecutive photos to give a view of the northern hemisphere of Jupiter from directly above the pole. Even though the visible appearance of the planet becomes less apparent at higher latitudes, discrete cloud features still seem to be zonally positioned. The spacings of features varies, and cloud systems have been seen to interact, one rolling over another. The broad white region is divided by the North Temperate Belt's high-speed jet stream, seen here as a thin brown line.

This page, below left: This is a photo, constructed from Voyager images, of the southern hemisphere of Jupiter from directly above the poles. This southern hemisphere image can be compared with its northern hemisphere companion by aligning the Great Red Spot. The southern hemisphere shows three white ovals, smaller scale spots, and the disturbance trailing from the Great Red Spot, which extends about 180 degrees in longitude from it.

Below: As this diagram clearly shows, the 16 known Jovian moons are organized into a very orderly system of four dissimilar groups (each comprised of four similar sized moons orbiting in distinctly different planes):

1. The inner group (except for Amalthea) were all discovered during the Voyager project, and they have diameters of less than 200 miles; they all orbit in a plane whose orbital inclination is less than half a degree and they are located less than 140,000 miles from Jupiter.

2. The second group, called 'The Galileans,' were all discovered in 1610 by Galileo and they have diameters greater than 1900 miles; they all orbit in a plane whose orbital inclination is less than half a degree and they are all between 250,000 and 700,000 miles from Jupiter.

3. The third group were all discovered in the twentieth century prior to Voyager, and they all have diameters of less than 105 miles; they all orbit in a plane whose orbital inclination is between 26 and 29 degrees and they are all between 6.9 and 7.2 million miles from Jupiter.

4. The final group were all discovered in the twentieth century prior to Voyager, and they all have diameters of less than 17 miles; they all orbit in a plane whose orbital inclination is between 147 and 163 degrees and they are all between 12.8 and 14.7 million miles from Jupiter.

the Voyager project. Adrastea is named for the mythical Greek king of Argos, whose daughter married Polynices of Thebes who had been exiled by his brother. Adrastea (or Adrastus) led several primitive expeditions against Thebes.

The second group of Jovian moons, the Galileans, were all discovered by Galileo during the first month (December 1609 —January 1610) of the great astronomer's telescopic survey of the heavens. Other than the Earth's Moon, they were the first planetary satellites to be observed and are easily observed from Earth with a telescope of moderate power. They are not only much much larger than any other Jovian moons, they are among the largest in the Solar System and Ganymede is *the* largest moon in the Solar System. They are discussed individually and in detail in the following pages.

The next group of Jovian moons are Leda, Himalia, Lysithea and Elara. Himalia and Elara were discovered by C D Perrine at Lick Observatory between November 1904 and February 1905. Lysithia was discovered by S B Nicholson at Lick Observatory in 1938, while Leda, discovered by Charles Kowal at Mount Palomar in 1974, was the last Jovian moon to be discovered from Earth. It is also *currently* the *smallest* known moon in the Solar System, though some observers think it may be larger than Deimos. Leda is named for the queen of Sparta who was the mother (by Zeus in the form of a swan) of Helen, Castor and Pollux.

The outermost group of Jovian moons, Ananke, Carme, Pasiphae and Sinope, are the moons most distant from their mother planets of any known in the Solar System. S B Nicholson discovered Ananke and Carme with the reflector telescope at Mount Wilson, California in September 1951 and July 1938 respectively. Pasiphae and Sinope were discovered by P J

Melotta at Greenwich, England in January 1908 and July 1914 respectively. Ananke is named for a Greek cult-goddess who shared a shrine at Corinth along with the goddess Bia. Horace later assigned her the Latin name Necessitas.

The four moons of the Jovian outer group also all orbit in a *retrograde motion.* The only other moons anywhere in the Solar System that move in retrograde motion are Saturn's Phoebe and Neptune's Triton.

Data for Adrastea (1979J1)
Diameter: 21.7 miles (35 km)
Distance from Jupiter: 83,080 miles (134,000 km)
Rotational period (Adrastian day): 7.11 Earth hours
Sidereal period (Adrastian year): 7.11 Earth hours

Data for Amalthea
Diameter: 168 × 103 × 93 miles (270 × 166 × 150 km)
Distance from Jupiter: 112,655 miles (181,300 km)
Rotational period (Amalthean day): 11.92 Earth hours
Sidereal period (Amalthean year): 11.92 Earth hours
Inclination to ecliptic plane: .4°

Data for Thebe (1979J2)
Diameter: 46.6 miles (75 km)
Distance from Jupiter: 137,690 miles (222,000 km)
Rotational period (1979J2 day): 16.23 Earth hours
Sidereal period (1979J2 year): 16.23 Earth hours

Data for Metis (1979J3)
Diameter: 24.9 miles (40 km)

Secondary Ring 44,267 mi
Primary Ring 76,136 mi
Metis (1979J3)
Adrastea (1979J1)
Amalthea
Thebe (1979J2)
Jupiter's cloud tops
Io
Europa
Ganymede
Callisto
Himalia
Leda
Lysith
76,136 mi   139,809 mi   664,867 mi   1,170,042 mi
6,897,220 mi   7,296,76

Distance from Jupiter: 79.28 miles (127,600 km)
Rotational period (1979J3 day): 7.08 Earth hours
Sidereal period (1979J3 year): 7.08 Earth days

Data for Leda
Diameter: 4.97 miles (8 km)*
Distance from Jupiter: 6,897,220 miles (11,100,000 km)
Rotational period (Ledan day): 238.7 Earth days
Sidereal period (Ledan year): 238.7 Earth days
Inclination to ecliptic plane: 26.7°
*Some data suggests that Leda may have a diameter nearly twice that which is given.

Data for Himalia
Diameter: 105 miles (170 km)
Distance from Jupiter: 7,127,127 miles (11,470,000 km)
Rotational period (Himalian day): 250.6 Earth days
Sidereal period (Himalian year): 250.6 Earth days
Inclination to ecliptic plane: 28°

Data for Lysithea
Diameter: 11.81 miles (19 km)
Distance from Jupiter: 7,276,256 miles (11,710,000 km)
Rotational period (Lysithean day): 259.2 Earth days
Sidereal period (Lysithean year): 259.2 Earth days
Inclination to ecliptic plane: 29°

Data for Elara
Diameter: 49.71 miles (80 km)
Distance from Jupiter: 7,296,761 miles (11,743,000 km)
Rotational period (Elaran day): 259.7 Earth days

Sidereal period (Elaran year): 259.7 Earth days
Inclination to ecliptic plane: 28°

Data for Ananke
Diameter: 10.56 miles (17 km)
Distance from Jupiter: 12,862,383 miles (20,700,000 km)
Rotational period (Anankean day): 617 Earth days
Sidereal period (Anankean year): 617 Earth days
Inclination to ecliptic plane: 147°

Data for Carme
Diameter: 14.9 miles (24 km)*
Distance from Jupiter: 13,887,646 miles (22,350,000 km)
Rotational period (Carmean day): 692 Earth days
Sidereal period (Carmean year): 692 Earth days
Inclination to ecliptic plane: 163°
*Some data suggests that Carme may have a diameter nearly twice that which is given.

Data for Pasiphae
Diameter: 16.78 miles (27 km)
Distance from Jupiter: 14,477,948 miles (23,300,000 km)
Rotational period (Pasiphaen day): 735 Earth days
Sidereal period (Pasiphaen year): 735 Earth days
Inclination to ecliptic plane: 148°

Data for Sinope
Diameter: 13.05 miles (21 km)
Distance from Jupiter: 14,694,000 miles (23,700,000 km)
Rotational period (Sinopean day): 758 Earth days
Sidereal period (Sinopean year): 758 Earth days
Inclination to ecliptic plane: 157°

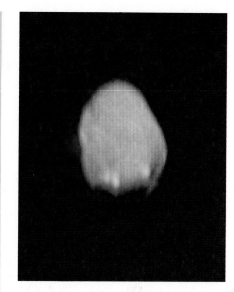

Above: Tiny, red Amalthea, Jupiter's innermost satellite, whizzes around the planet every 12 hours, only 1.55 Jupiter radii from the cloud tops. In this view taken by Voyager 1 from a range of 255,000 miles on 4 March 1987, the satellite appears to be about 80 miles high by 100 miles wide. Part of its longer dimension is not illuminated. North is at right, Jupiter is to the top. The reflectivity of the surface is less than 10 percent, making Amalthea much darker than the Galilean satellites. Amalthea's irregular shape probably results from a long history of impact cratering. Some of the indentations near the bottom and at upper right may be marginally resolved craters. This irregular satellite probably keeps its long axis pointed toward Jupiter in its motion around the planet so that the spin period around its own axis equals its period of revolution around Jupiter (12 hours). Amalthea was discovered in 1892 by the American astronomer Edmund Emerson Barnard at Lick Observatory.

Note: The specifications above cover all the known Jovian moons except the four Galileans. Data for Io, Europa, Ganymede and Callisto are given on the following pages along with maps and other more detailed information.

Ananke
12,862,383 mi

Carme
13,887,646 mi

Sinope
Pasiphae
14,694,000 mi

IO

The innermost of Jupiter's Galilean moons, Io, is one of the most intriguing bodies in the Solar System. In 1979, the photographs returned by the Voyager spacecraft revealed a very dynamic world whose surprising characteristics were beyond anything that had previously been imagined. The most volcanically active body in the Solar System, Io is the only place besides Earth where volcanic eruptions have actually been observed.

Named for the maiden in Greek mythology who became a lover of Zeus only to be turned into a cow by Hera, the moon called Io is a unique world. Io's surface, with its brilliant reds and yellows that remind one of a giant celestial pizza, is a crust of solid sulfur 12 miles thick that floats on a sea of molten sulfur. This sea is in turn thought to cover silicate rock which may be partially solid but is at least partially molten. The tidal effect of so massive a body as nearby Jupiter is thought to cause the crust to rise and fall on the molten sulfur by as much as 60 miles. It is through this heaving sulfur crust that Io's volcanos have burst.

The red and yellow sulfurous surface is marred by dozens of jet black volcanos whose violent eruptions surpass (both in magnitude and frequency) anything seen on Earth. During its brief encounter with Io in March 1979, Voyager 1 observed the erup-

Data for Io

Diameter: 2257 miles (3632 km)
Distance from Jupiter: 261,970.09 miles (421,600 km)
Mass: 1.97×10^{23} lb (8.92×10^{22} kg)
Rotational period (Ioan day): 1.76 Earth days
Sidereal period (Ioan year): 1.76 Earth days
Inclination to ecliptic plane: 0°
Mean surface temperature: −243° F
Largest known surface feature: Pele (620 miles wide)

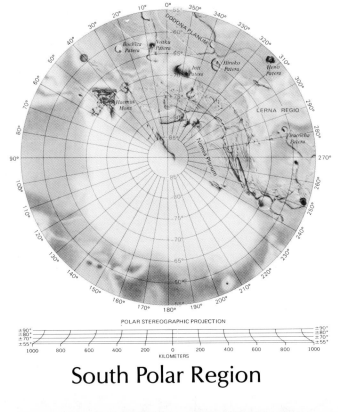

POLAR STEREOGRAPHIC PROJECTION

South Polar Region

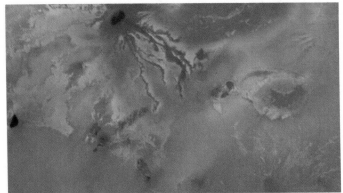

Above: Io's Ra Patera volcano with sulfurous lava flows in evidence.

MERCATOR PROJECTION

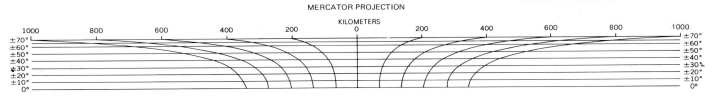

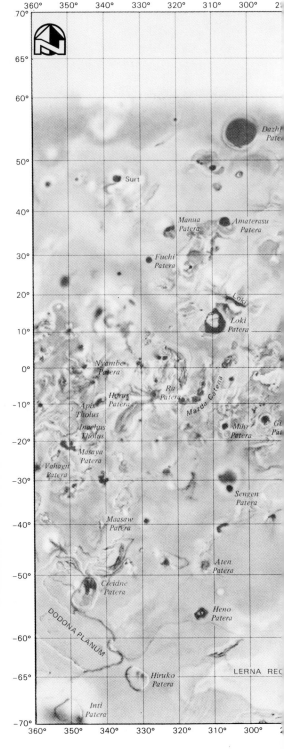

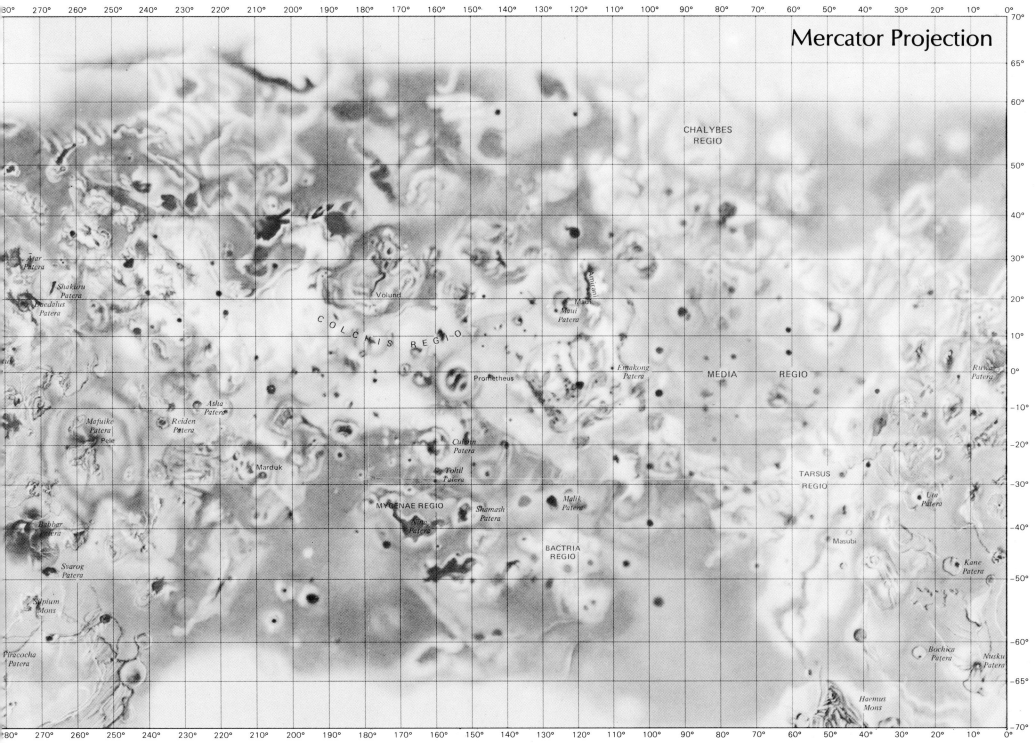

Mercator Projection

COLCHIS REGIO

CHALYBES REGIO

MEDIA REGIO

TARSUS REGIO

BACTRIA REGIO

MYCENAE REGIO

Ivar Patera

Shakuru Patera

Daedalus Patera

Volund

Amirani

Maui

Maui Patera

Emakong Patera

Prometheus

Ruwa Patera

Asha Patera

Mafuike Patera

Reiden Patera

Pele

Culann Patera

Uta Patera

Marduk

Tohil Patera

Malik Patera

Babbar Patera

Shamash Patera

Masubi

Sigurd Patera

Kane Patera

Svarog Patera

Silpium Mons

Pillan Patera

Bochica Patera

Viracocha Patera

Nusku Patera

Haemus Mons

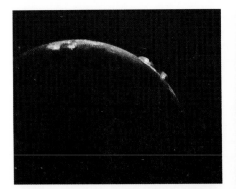

At top: Voyager 2 took this picture of Io, on the evening of 9 July 1979. The two blue volcanic eruption plumes seen here originate from the volcanoes Amirani (upper) and Maui (lower).

Above: A color reconstruction of one of the erupting volcanos on Io discovered by Voyager 1 on 4 March 1983. In this color analysis the region that is brighter in ultraviolet light (blue in this image) is much more extensive than the denser, bright yellow region near the center of the eruption.

Above right: This image of Io was made from several frames taken by Voyager 1 on 4 March 1979. The circular, donut-shaped feature in the center is the active volcano Prometheus. Io is the first body in the solar system (beyond Earth) where active volcanism has been observed.

tions of no fewer than eight volcanoes, with the plume above the volcano 'Pele' reaching an altitude of 174 miles. When Voyager 2 turned its cameras toward Io in July 1979, it was able to observe all the volcanoes discovered by Voyager 1 except Volund. Of these, only Pele showed no continuing volcanic activity. Amirani, Marduk, Masubi and Maui displayed activity similar to that observed in March, while Prometheus and Loki were more violent than before. The plume above Loki had become a double plume and had more than doubled in altitude from 60 to 130 miles! There were in fact no *small* eruptions recorded among active volcanos on either pass, as the smallest plumes all exceeded an altitude of 40 miles.

In these violent eruptions, sulfurous material was belched from Io's liquid mantle at speeds of up to 3280 feet per second —many times the recorded velocity of the Earth's volcanoes. The reason for the extreme altitudes of the plumes and the high velocity of the particles is due in part to the weaker gravity on Io—whose mass is less than two percent that of the Earth. The sulfur particles fall to the surface relatively fast, however, because there is no atmosphere on Io in which they could become suspended, and hence no winds to blow them in great billowing clouds of ash across the land—as was the case on Earth following the eruptions of Mount Etna or Mount St Helens.

Each of Io's eruptions dumps 10,000 tons of sulfur onto the moon's surface. In extrapolation, this would account for 100 billion tons of sulfur deposits per year. This is enough to cover the entire surface with a layer of sulfur 'ash' a foot thick in 30,750 years. Combined with surface flows, Io could very well be completely resurfaced with a foot-thick layer in as short as 3100 years, giving this pizza-colored moon the youngest solid surface in the Solar System aside from Earth, and there are parts of Earth that change less over time than much of Io. In fact, there were many noticeable changes in Io's surface—particularly around Pele—in just the four months between March and July 1979. This 'ever-youthful' surface accounts for the complete absence of meteorite impact craters on Io.

Surrounding the black volcanic caldera are black fan-shaped features that are the result of liquid sulfur cooling rapidly as it reaches Io's frigid surface. South of Loki, the Voyager imaging team discovered a U-shaped molten sulfur lake 125 miles across that had partially crusted over. It was detected by its surface crust temperature of about 65 degrees Fahrenheit—compared to the surrounding surface temperature of less than −230 degrees. This lake has certainly cooled and solidified by now, while other molten sulfur lakes have no doubt formed elsewhere.

Io's volcanic activity appears to be of at least two kinds—explosive eruptions (*above*) that spew material as high as 160 miles above the moon's surface, and lava flows that smolder and gurgle across the pizza-like face of Io.

Another common product of Io's eruptions is sulfur dioxide. In the atmospheres of Venus and Earth, sulfur dioxide gas mixes with water to form sulfuric acid. On Io, there is no water and it is so cold that sulfur dioxide exists as the fine white snow-like solid which is seen to dust various regions, or form rings around some of the larger volcanic mountains.

While Io has no atmosphere in the usual sense, there is a donut-shaped *torus* or tube of electrically charged particles that exists in the path of its orbit around Jupiter. The torus apparently originates in the material from the eruptions, and consists of ionized sulfur and oxygen atoms.

EUROPA

With a surface that is probably composed entirely of water ice, and which is marred only by three definite impact craters, Europa is an enigma. The absence of impact craters on Io is explained by its violently active surface, but Europa has an extremely smooth and apparently inactive surface. Named for the Phoenician princess abducted to Crete by Zeus, Europa has a highly reflective surface that probably remained in a slushy semi-liquid state until relatively recently. This is at present the only explanation for its lack of meteorite craters.

The smooth surface is, however, not without features and these too present part of Europa's mystery. The features include long black *linea,* or lines reminiscent of Percival Lowell's 'canals' on Mars. These features, which are up to 40 miles wide and stretch for thousands of miles across the surface, defy explanation. They appear to be cracks in the lighter surface, but they have no depth and thus they can only be described as 'marks' on the surface.

Another peculiar feature of Europa's icy surface are numerous dark *macula,* or spots distributed across the surface. While the three impact craters range from 11 to 15.5 miles across, the macula are generally smaller than six miles.

The most unusual features on Europa are the *flexus.* Light colored, scalloped ridges, the flexus are much narrower and somewhat shorter than the linea. They are, however, more regular in width and in the regularity of their scallops or cusps.

Europa is thought to have a relatively large silicate core with a layer of molten silicate above that which is in turn covered by a layer of liquid water perhaps 60 miles deep. Above this is the strange icy crust which is roughly 40 miles thick.

Data for Europa

Diameter: 1942 miles (3126 km)
Distance from Jupiter: 416,877 miles (670,900 km)
Mass: 5.25×10^{23} lb (4.87×10^{22} kg)
Rotational period (Europan day): 3.55 Earth days
Sidereal period (Europan year): 3.55 Earth days
Inclination to elliptic plane: .5°

Below left: Europa, the brightest of Jupiter's Galilean satellites, appears to have a thin smooth crust of ice overlying water or soft ice. The complex patterns on its surface suggest that the icy surface is fractured and that the cracks are filled with dark material from below. Very few impact craters are visible on the surface, perhaps indicating that active processes of the surface are still modifying Europa. The surface patterns of Europa differ drastically from the patterns and fault systems seen on Ganymede, where pieces of the crust have moved in relation to one another. Europa's crust evidently fractures, but the pieces remain approximately in their original positions.

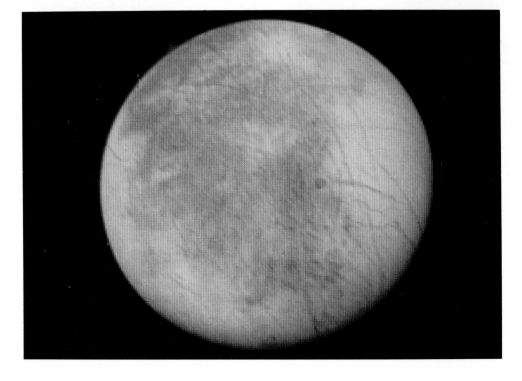

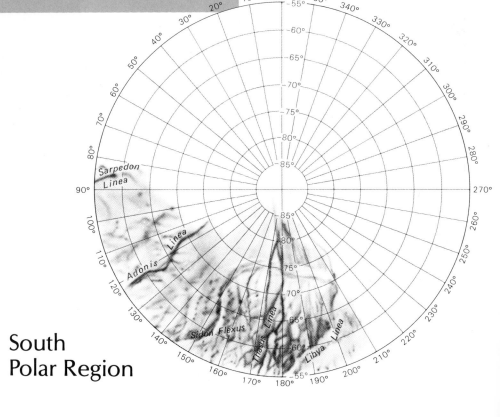

South
Polar Region

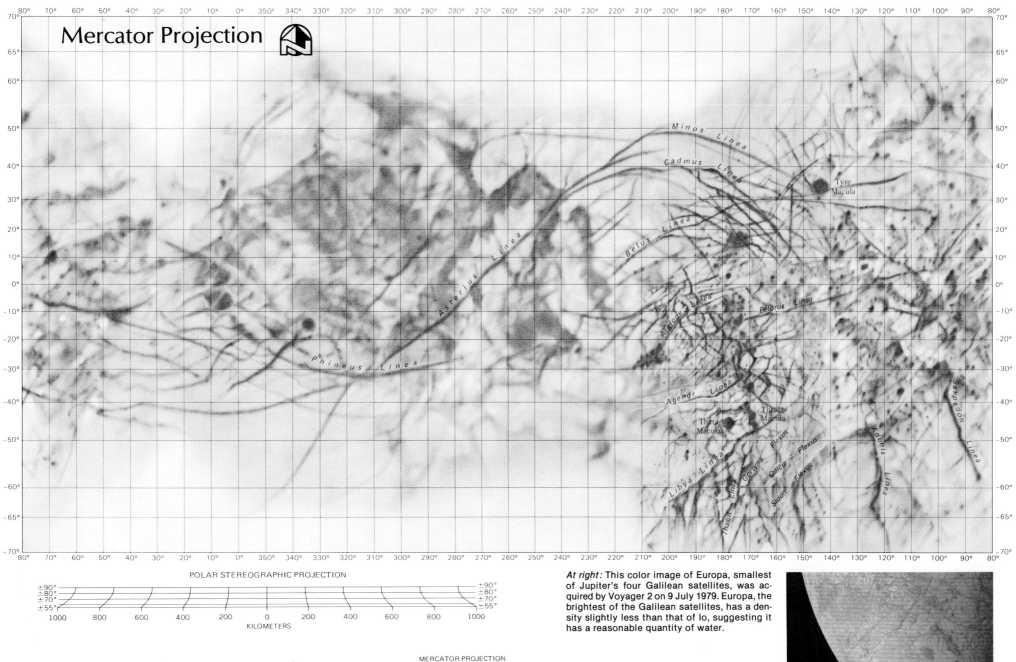

Mercator Projection

Minos Linea
Cadmus Linea
Tyre Macula
Belus Linea
Asterius Linea
Araiode Linea
Pelorus Linea
Phineus Linea
Agenor Linea
Thrace Macula
Thera Macula
Libya Linea
Cortyna Flexus
Cilicia Flexus
Sidon Flexus
Sarpedon Linea
Adonis Linea
Thasus Linea

POLAR STEREOGRAPHIC PROJECTION

±90°
±80°
±70°
±55°

1000 800 600 400 200 0 200 400 600 800 1000

KILOMETERS

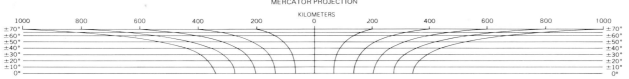

MERCATOR PROJECTION

KILOMETERS

1000 800 600 400 200 0 200 400 600 800 1000

±70°
±60°
±50°
±40°
±30°
±20°
±10°
0°

At right: This color image of Europa, smallest of Jupiter's four Galilean satellites, was acquired by Voyager 2 on 9 July 1979. Europa, the brightest of the Galilean satellites, has a density slightly less than that of Io, suggesting it has a reasonable quantity of water.

GANYMEDE

Named for the cup-bearer of the Greek gods, Ganymede is the largest moon in the Jovian system and is, indeed, the largest moon in the entire Solar System. Ganymede, like Callisto, is composed of silicate rock and water ice, and thus these bodies came to be dubbed 'dirty snowballs.' Ganymede has an ice crust that is roughly 60 miles thick. This crust in turn floats upon a mantle of slushy water that is roughly 400 miles deep. Beneath Ganymede's mantle is a heavy silicate core.

Like Earth's rocky crust, Ganymede's icy crust is divided into plates which shift and move independently, interacting with one another along fracture zones, resulting in geologic activity that is very much like that which has been observed on Earth. Mountain ranges 10 miles wide and 3000 feet high have formed on Ganymede as a result of the pressure of ice plates against one another. The surface of Ganymede is characterized by mountainous terrain and ancient dark plains, the largest of which is the region known as Galileo Regio. The dark plains are in turn marked by a wrinkled or grooved terrain consisting of a semi-circular system of parallel curved ridges six miles wide, 325 feet high and approximately 40 miles apart. These grooves are the remnants of an ancient impact basin that has long since been obscured by subsequent geologic activity.

A great number of smaller impact craters have been identified on Ganymede's surface, with many of them showing white 'halos.' These are evidence of water having been splashed up through the crust after each meteorite smashed its way through the surface.

Because of the size of Ganymede and the presence of liquid water, it once was suggested that there might be a tenuous atmosphere of water vapor and free oxygen (with the latter being formed by the effect of sunlight on the former). No evidence of an atmosphere was detected by the Voyager spacecraft in 1979, and indeed if one were present it would have to have an atmospheric pressure less than one hundred billionth of Earth's.

Data for Ganymede

Diameter: 3278 miles (5276 km)
Distance from Jupiter: 664,867 miles (1,070,000 km)
Mass: 3.29×10^{23} lb (1.49×10^{23} kg)
Rotational period (Ganymedian day): 7.16 Earth days
Sidereal period (Ganymedian year): 7.16 Earth days
Inclination to ecliptic plane: .2°

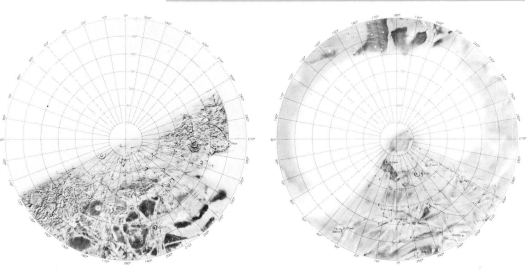

Above: Voyager 2 took this picture of Ganymede. The boundary of the largest region of dark, ancient terrain can be seen to the right.

North Polar Region   South Polar Region

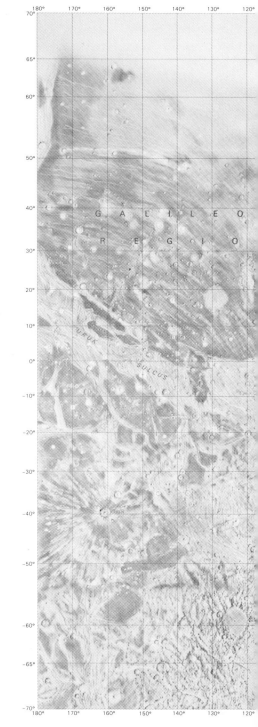

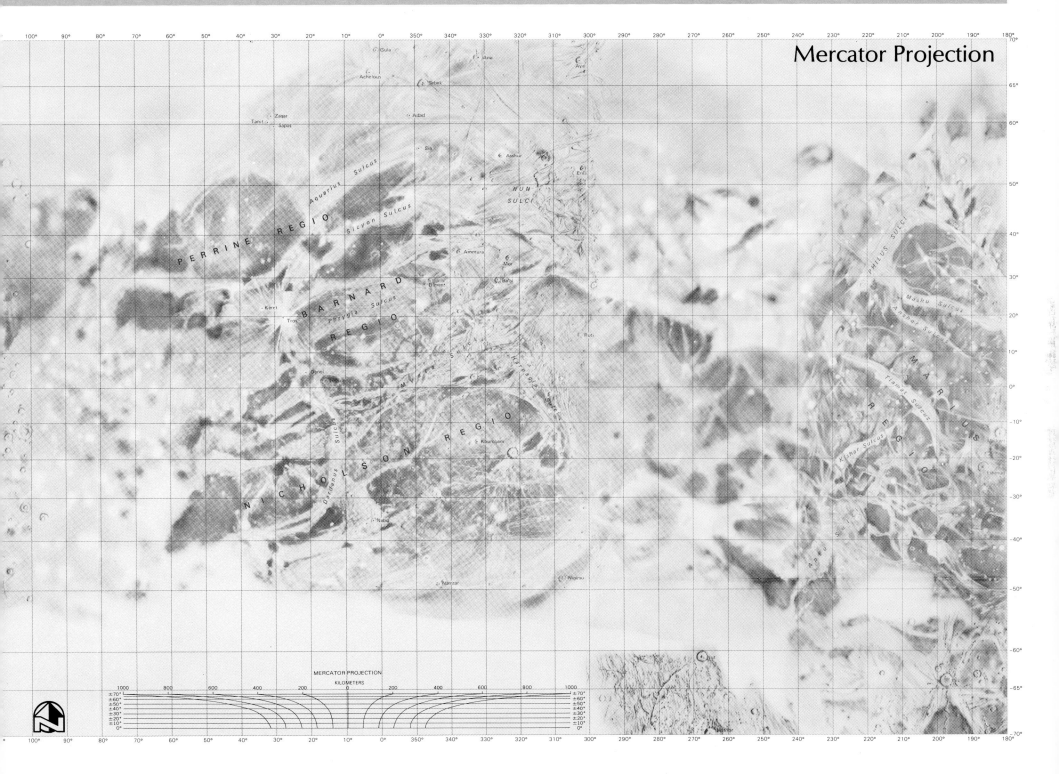

Mercator Projection

MERCATOR PROJECTION
KILOMETERS

CALLISTO

Named for a Greek nymph favored by Zeus and turned into a bear by the jealous Hera, Callisto has a water ice and silicate rock composition like that of Ganymede. Unlike Ganymede, Callisto's ice and silicate soil surface shows no sign of any geologic activity. The only surface feature on this 'dirty snowball' is a mass of hundreds upon hundreds of impact craters. The largest of these is Valhalla, a huge impact basin in Callisto's northern hemisphere that measures 1860 miles across.

The probable reason for the lack of geologic activity is that Callisto's icy crust is more than 150 miles thick, and thus is not prone to break into plates as Ganymede's has. It is also much farther from the tidal effects of Jupiter's gravity. Beneath the solid ice crust is a slushy mantle 600 miles in depth, and beneath the mantle, a heavy silicate core. Thus, Callisto is thought to be identical to Ganymede in terms of its composition, but with a thicker and hence more geologically inert crust.

With temperatures ranging between −200 and −300 degrees Fahrenheit, there is little likelihood of an atmosphere hanging over Callisto's frigid wastes.

Data for Callisto

Diameter: 2995 miles (4820 km)
Distance from Jupiter: 1,170,041 miles (1,883,000 km)
Mass: 2.35×10^{23} lb (1.06×10^{23} kg)
Rotational period (Callistan day): 16.689 Earth days
Sidereal period (Callistan year): 16.689 Earth days
Inclination to ecliptic plane: .2°
Mean surface temperature: −279 F
Largest surface feature: Valhalla (a circular basin)
1864 miles (3000 km) across

Above: Callisto's icy, dirt-laden surface appears to be very ancient and heavily cratered. Large concentric rings are evidence of several enormous impacts caused by huge meteors crashing into the surface.

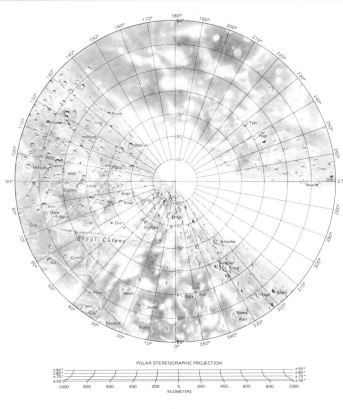

POLAR STEREOGRAPHIC PROJECTION

North Polar Region

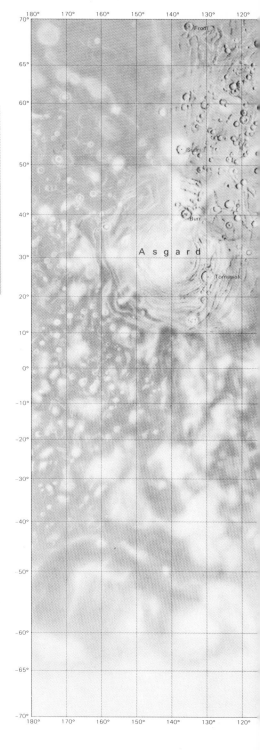

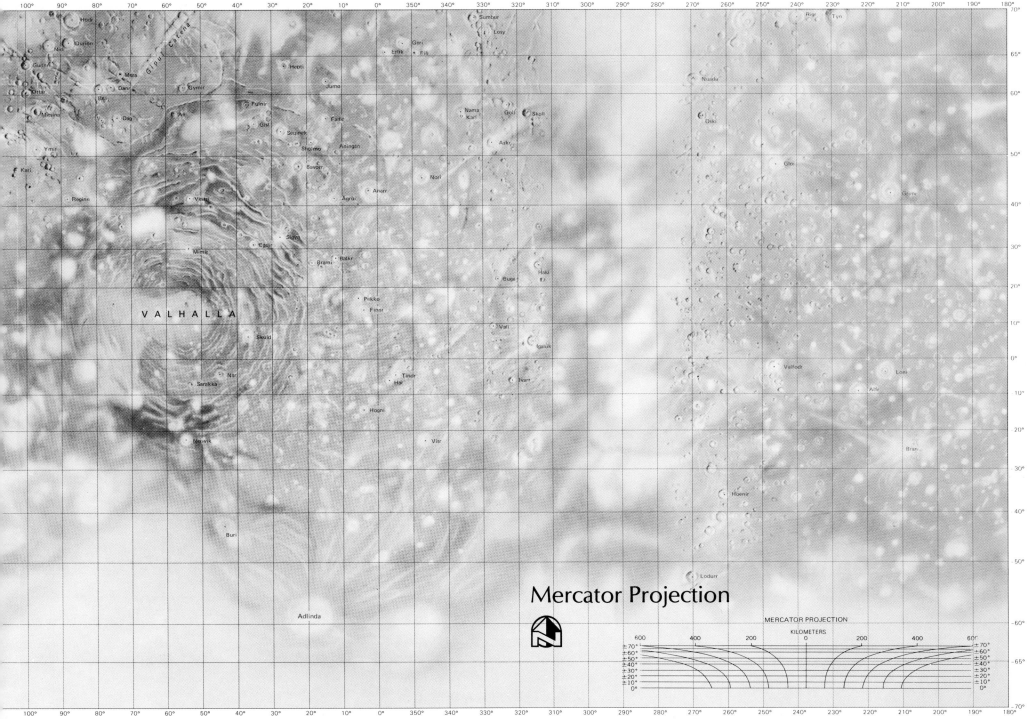

Mercator Projection

SATURN

There is no object in orbit around the Sun whose appearance is more awesome and spectacular than the planet Saturn. In terms of sheer size, Saturn dwarfs all the other planets except Jupiter, but its incredible system of *rings* puts it visually in a class by itself.

Named for Jupiter's father, the original patriarch of Roman gods, Saturn is the outermost of the planets visible from Earth with the unaided eye.

Saturn's composition is very much like that of Jupiter. There is probably a solid *core* composed of iron and silicates that measures less than 9300 miles in diameter and is covered by a layer of water kept in solid ice form by the pressure of successive higher layers of metallic hydrogen and liquid molecular hydrogen. Like Jupiter, nearly 80 percent of Saturn's mass is hydrogen, the simplest element, with most of the remainder taken up by helium, the second simplest and second most common element in the Solar System. Thus Saturn may have had the same star-like ancestry as Jupiter. Major components of the core are iron (0.2 percent) and silicates (0.1 percent) as well as oxygen (1 percent) which is also present, with hydrogen, in the water ice. The remaining element is the inert gas neon (0.2 percent) which, along with organic gases composed of nitrogen (0.1 percent) and carbon (0.4 percent), comprise Saturn's atmosphere.

Saturn's *atmosphere* contains many of the same gases that are present in the atmospheres of Jupiter, Uranus and Neptune. The most common are methane and ammonia, but there are also trace amounts of phosphene and such more complex organic compounds as propane, ethane, acetylene and methylacetylene. Saturn's atmosphere is a good deal smoother, hazier and less choppy than Jupiter's, with relatively few distinct features to parallel the Jovian brown, white and Great Red spots. Saturn's atmosphere is characterized by horizontal bands alternating between those with westerly, or the rarer easterly, prevailing winds. These winds have speeds of up to 900 mph, with the greatest westerly velocities being recorded within five degrees latitude of the equator. The largest feature in Saturn's atmosphere is Anne's Spot, a pale red feature in the southern hemisphere that is similar to, though smaller than, Jupiter's Great Red Spot. Like the latter, Anne's Spot is thought to be composed of phosphine that is brought high into the upper atmosphere by spiralling convection currents.

The fact that Saturn's cloud tops are smoother than Jupiter's may be due to weaker gravity (because of smaller mass) and lower temperatures. At these colder temperatures the condensation point of the chemicals in the atmosphere would be reached in regions of higher pressure and hence at lower altitudes within the atmosphere. Thus Saturn's atmosphere is probably characterized by the same visible turbulence as Jupiter's, but at lower altitudes below a layer of ammonia haze.

Saturn's ring system, while not absolutely unique, is certainly the planet's outstanding feature. Galileo Galilei (1564–1642) first observed the rings in 1610, but Saturn happened to be oriented so that the great Italian astronomer was viewing them nearly edge-on and thus it wasn't clear what they were. Galileo

Facing page: This awe-inspiring view of Saturn was assembled from photographs taken by the American Voyager 2 spacecraft on 4 August 1981 from a distance of 13 million miles on the spacecraft's approach trajectory. Three of Saturn's icy moons are evident here. They are, in order of distance from the planet: Tethys, (652 miles in diameter), Dione (696 miles) and Rhea (951 miles). The shadow of Tethys appears in this photo on Saturn's southern hemisphere. A fourth satellite, Mimas, is less evident, appearing as a bright spot near the planet's limb above Tethys. The shadow of Mimas appears on the planet directly above that of Tethys. The pastel and yellow hues on the planet reveal many constrasting bright and darker bands in both hemispheres of Saturn's weather system.

Above: Saturn as viewed through the 60 inch telescope at Hale Observatory. Saturn's rings can be seen from Earth with even a moderately powered telescope.

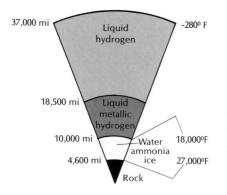

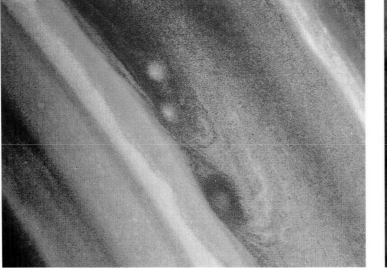

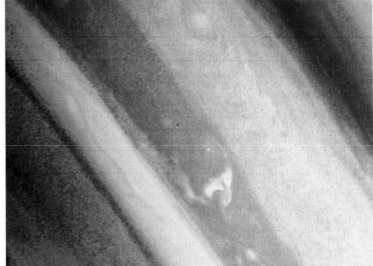

Usually Saturn's rings get all the attention. On these pages, we give some consideration to the atmosphere of the planet itself. *Above* is a simple pie slice graph illustrating Saturn's composition from its rocky core to the upper reaches of its dense atmosphere.

At immediate upper right: This photo of Saturn's northern hemisphere is a false color image constructed from Voyager 2 data. Note the blue atmospheric vortice in the center and the two vortices below and to the right of it. The striations are caused by latitudinally oriented winds, which reach velocities of up to 900mph.

At far upper right is a similarly enhanced image of some of Saturn's clouds, which are actually swirls or eddies of atmospheric material.

Facing page: The convective clouds between the two spots are typical of this region. Seen above the planet is the moon Enceladus.

At right: This Voyager 2 spacecraft observation of Saturn's northern mid-latitudes reveals a strangely curled cloud attached by a thin ribbon to the bright white cloud region in mid-photo. The cloud was monitored for seven rotations around the planet, and appeared to be forming a closed loop.

at first thought he had discovered two identical moons of the scale that he had found near Jupiter. However, these 'moons' did not rotate or change position and Galileo was mystified. He wrote to the Grand Duke of Tuscany that 'Saturn is not alone but is composed of three, which almost touch one another and never move nor change with respect to one another. They are arranged in a line parallel to the zodiac, and the middle one *(Saturn itself)* is about three times the size of the lateral ones' *(actually the outer edges of the rings).*

By 1612, the plane of the rings was oriented *directly* at the Earth and the 'lateral moons' seemed to disappear entirely. Galileo was completely baffled, but no less so than when they reappeared in 1613. In December 1612, Galileo had written 'I do not know what to say in a case so surprising, so unlooked for and so novel. The shortness of the time, the unexpected nature of the event, the weakness of my understanding, and the fear of being mistaken have greatly confounded me.'

Data for Saturn

Diameter: 74,565 miles (120,000 km)
Distance from Sun: 934,340,000 miles (1,507,000,000 km)
at aphelion
835,140,000 miles (1,347,000,000 km)
at perihelion
Mass: 2.5836×10^{26} lb (5.684×10^{26} kg)
Rotational period (Saturnian day): 10.25 Earth hours
Sidereal period (Saturnian year): 10,759 Earth days
(29.46 Earth years)
Eccentricity: 0.056
Inclination of rotational axis: 26.73°
Inclination to ecliptic plane (Earth = 0): 2.49°
Albedo (100% reflection of light = 1): .33
Mean temperature: −284.8° F
Maximum temperature: −207.4° F
Largest feature: Anne's Spot (a large red spot)
3107 × 1864 miles
Major atmospheric components: Hydrogen (94%)
Helium (6%)
Other atmospheric components: Ammonia, Phosphine,
Methane, Ethane, Methylacetylene,
Acetylene, Propane
Atmospheric depth: 90 miles

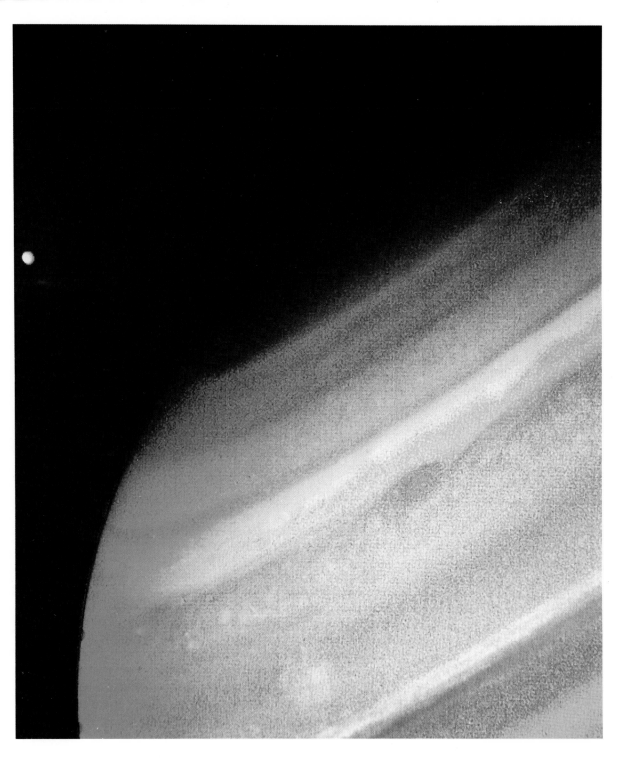

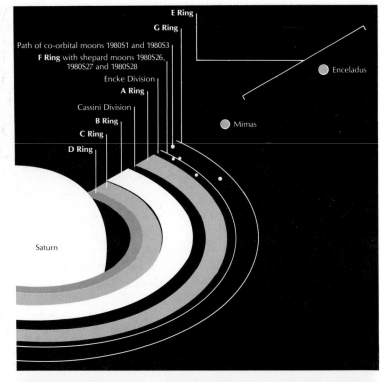

At top above: These drawings from *Systema Saturnium* (published in 1659) were made before the true nature of Saturn's rings was understood. In the early seventeenth century, astronomers had no concept of a ringed planet, so it was postulated that Saturn might be accompanied by a pair of huge moons. In the illustration here, drawing I was based on observations made by Galileo in 1610; II was made by Scheiner in 1614; and III was made by Riccio in 1641.

As telescopes improved, so the understanding of Saturn's rings improved. The *above* drawing was made by JD Cassini—this is the first representation of the Cassini Division (which was subsequently named for the astronomer) between the A and B Rings.

Above right: This is our modern understanding of Saturn's system of rings and moons.

At right: This Voyager 1 photo clearly reveals a startling discovery—the 'braided' structure of the F Ring, which appears to be two narrow bright rings twisted together, with a dimmer companion ring just slightly 'inboard' of them. Note the 'kink' in this ring system near the top of the photo.

In 1655, more than a decade after Galileo's death, the Dutch astronomer Christiaan Huygens (1629–1695) solved the riddle. Using a telescope more powerful than that which was available to his predecessor, Huygens figured out that the mysterious objects were rings around Saturn; and the reason for their 'disappearance' in 1612. He also went on to calculate that the rings would be oriented in this way on a 150-month cycle, and that at opposing ends of the same cycle almost the entire ring would be visible from Earth. It has since been determined that the cycle actually alternates between periods of 189 and 165 months. Huygens also discovered Saturn's largest moon, Titan.

In 1671, the Italian-born and naturalized French astronomer Giovanni Domenico (aka Jean Dominique) Cassini (1625–1712) began his own observations of the ringed planet. Cassini discovered a second moon of Saturn, Iapetus, in 1671 and in 1675 he determined that the 'ring' around Saturn was not a single band, but a pair of concentric rings. These two rings would come to be known as the A Ring and B Ring, with the space between them appropriately named the Cassini Division.

In 1837, Johann Franz Encke (1791–1865), at the Berlin Observatory, tentatively identified a faint division in the A Ring. This division was confirmed in 1888 by James Keeler (1857–1900) of Allegheny Observatory in the United States.

Subsequently this division is known as either the Keeler Gap or (more often) as the Encke Division.

The first spacecraft to venture close to Saturn was the American Pioneer 11 in September 1979. Prior to this time, there were only three known rings of Saturn, each lettered in the order of their discovery from A through C. Pioneer 11 helped Earth-based astronomers to identify a fourth ring which is now known as the F Ring. When the two American Voyager spacecraft first approached Saturn in November 1980, the spectacular photographs they beamed back to Earth revealed that there were not just four, six or even a dozen rings in Saturn's ring system; rather, there were literally thousands of rings, with each known ring itself composed of hundreds or thousands of rings, with faint rings even identified within the Cassini Division.

In the new nomenclature of Saturn's ring system, the thousands of now known rings and ringlets are divided into seven main rings based on the older nomenclature. Closest to Saturn is a wide but sparse and extremely faint ring known as the D Ring. At a point roughly 46,000 miles about Saturn's center and 8700 miles above Saturn's cloud tops, the D Ring merges into the more distinct C Ring. The C Ring is 10,850 miles wide, making it the second widest of the easily-visible main rings though it is less prominent than the slightly narrower A Ring.

At a point 19,600 miles above Saturn's cloud tops, the C Ring merges into the B Ring without a major gap. The B Ring is 15,800 miles wide and is the brightest of Saturn's main rings and the widest, except for the virtually invisible E Ring. The 2800 mile-wide Cassini Division separates the B Ring from the 9000 mile-wide A Ring, the second brightest of Saturn's rings.

The bright yet tenuous F Ring is located just 2300 miles beyond the outer edge of the A Ring and the narrow G Ring is located 20,700 miles from the A Ring. The E Ring is a very, very faint 55,800 mile-wide mass of particles that begins 91,000 miles

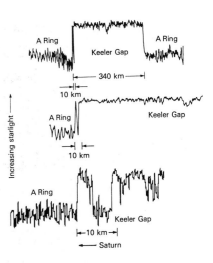

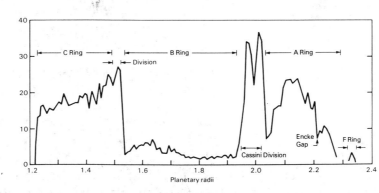

Facing page, lower right: The dark spokes in the B Ring of Saturn can be seen in this Voyager 2 photo. After the spacecraft had passed the planet and viewed these spokes in forward-scattered light, the spokes appeared as bright streaks, indicating that they are caused by tiny particles no larger than the wavelength of light.

Below left: The blue color of the C Ring and the Cassini Division as well as the many color variations within the B Ring and the A Ring are due to special computer color recording techniques, and are useful, in this heightened mode, to identify possible variations in chemical composition within the bands of the ring system.

Above and above left : These are brightness graphs of Saturn's ring system — above left represents ring brightness as seen from the Ring System's unilluminated side, and above represents the brightness of a single ring as seen from the illuminated side—note that the modalities of this graph are inversions of those shown at above left. The graph above also measures ring brightness as compared to background light—in this case, starlight—which increases as one reads from the bottom up.

from Saturn's cloud tops and extends past the orbit of the moon Enceladus.

Saturn's rings are composed of silica rock, iron oxide and ice particles which range from the size of a speck of dust to the size of a small automobile. In density they range from the nearly opaque B Ring to the translucent E Ring.

Theories about the Ring system's origin generally fall into two camps. One theory, originating with Edward Roche in the nineteenth century, holds that the rings were once part of a large moon whose orbit decayed until it came so close to Saturn as to be pulled apart by the planet's tidal or gravitational force. An alternate to this theory suggests that a primordial moon disintegrated as a result of being struck by a large comet or meteorite. One opposing theory is that the rings were never part of a larger body, but rather they are nebular material left over from Saturn's formation 4.6 billion years ago; in other words they were part of the same pool of material out of which Saturn formed, but they remained separate and gradually formed into rings.

Whatever their origin, however, Saturn's rings will continue to distinguish the planet and make it what Galileo described in 1610 as a 'most extraordinary marvel.'

The Saturnian Ring System

	Ring width (in miles)	Orbital Distance (Planet center to inner edge of ring in miles)	Ring Composition
D ring	4588	41,540	dust-like particles
C ring	10,850	45,384	water ice
B ring	15,810	56,854	water ice
A ring	9114	75,020	water ice
F ring	62(?)	87,172	
G ring	75(?)	105,400	dust-like particles
E ring	29,760	112,220	

Successful Earth Expeditions to Saturn

Name	Country of Origin	Launch	Date of closest contact or landing
Pioneer 11	USA	5 Apr 73	1 Sep 79
Voyager 2	USA	20 Aug 77	25 Aug 81
Voyager 1	USA	5 Sep 77	12 Nov 80

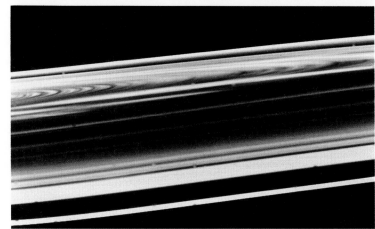

Facing page: The large, bright spots in Saturn's North Temperate Belt might closely resemble gigantic convective storms, much larger than thunderstorms on Earth, upwelling from deep within Saturn's atmosphere. The largest violet-colored cloud belt in this false-color image is Saturn's North Equatorial Belt.

The Southern Hemisphere of the planet (below the rings) appears bluer than the Northern Hemisphere because of the increased scattering of sunlight upon that area due to the spacecraft's point-of-view.

Above: The Encke Division in Saturn's outer A Ring is shown in this computer-enhanced photograph.

At left: The Saturn ring system, as viewed across the plane of the rings, with Saturn off-camera to the left.

SATURN'S MOONS

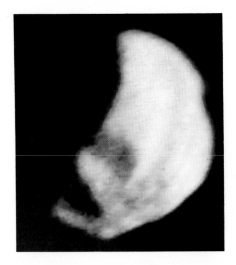

Two satellites of Saturn share an orbit; the trailing co-orbital satellite, designated 1980 S3, but sinced named Epimetheus, is seen in the image *above*. The shadow of Saturn's F-ring was photographed transitting the moon; the colored stripes seen on the moon are the shadow's image, photographed through variably colored filters.

Below, these pages: A diagram of Saturn's moon system, showing their relative distances from the mother planet and her ring system. The moons are drawn to scale with one another, but are exaggerated in relation and distances. Co-orbital moons are shown close together for the sake of clarity.

Not only do its spectacular rings set Saturn apart from other planets, but so too does its complex system of more than 20 moons. Saturn's moons range in size from huge Titan, once thought to be the Solar System's largest moon, to the family of tiny moons that were discovered in photographs taken by the Voyager spacecraft in 1980. Though the moons of Saturn are no less diverse in character than those of Saturn, they are generally smaller and, with the exception of the two outermost (Iapetus and Phoebe), their orbital inclination is within 1.5 degrees of that of the rings. With the exception of Phoebe, the moons are synchronous, like Earth's Moon, meaning that the same side faces Saturn at all times. The western hemispheres, which face in the direction of their orbital paths, are called *leading hemispheres* while the eastern are called *trailing hemispheres*.

In 1977 when the two Voyager spacecraft were launched from Earth, the ringed planet was known to have nine moons. Five of these were discovered prior to 1700 (with four found by Giovanni Cassini, the man who discovered that Saturn had multiple rings); and only two were discovered after 1800. By the time that the Voyager data was digested in 1982, Saturn was known to have 17 moons and four to six additional Lagrangian *co-orbital satellites*. A co-orbital is one of a group of moons that share a single orbital path, while Lagrangian satellites (named after the eighteenth century astronomer whose mathematical theory postulated their existence) are small co-orbitals that exist in the orbit of a larger moon 60 degrees ahead or 60 degrees behind it in the orbital path.

Stephen Synnott of NASA's Jet Propulsion Laboratory has led the way in the identification of Lagrangian moons and in

postulating the existence of others. Such moons have been found sharing the orbits of Mimas and Dione and *four* such moons have been found with Tethys. There is also the possibility that Dione has a second co-orbital and that Tethys has a fifth.

Saturn's most recently discovered moons are so small and so close to the planet that it is almost hard to know where to draw the line between moons and ring particles. This also makes it harder to find such bodies visually against the brilliant rings.

The innermost of Saturn's moons is Atlas (originally 1980 S28) which is named for one of the Titans of Greek mythology who was condemned to support the weight of the universe on his shoulders. Atlas is also informally known as the A Ring Shepherd Moon because of its role in shepherding the nearby outer A Ring particles and in a sense *defining* the outer edge of the A Ring.

The next two moons, Prometheus and Pandora (originally 1980 S27 and 1980 S26) are named respectively for the Titans of Greek mythology who stole fire from Olympus to give it to man, and for the woman who was bestowed upon man as a punishment for Prometheus having stolen fire. Prometheus and Pandora are known informally as F Ring Shepherd Moons because of their positions on either side of the F Ring and their role in defining that ring. The two moons may also be responsible for the kinks and braiding observed in the F Ring.

Well beyond the F Ring but inside of the G Ring are Epimetheus and Janus (originally 1980 S3 and 1980 S1) which are the first of the several groups of co-orbital moons and the only group of co-orbitals not to be of the Lagrangian type. The centers of these two bodies and hence the 'center lines' of their orbital paths are offset by only 30 miles, a distance narrower than the radius of either! Epimetheus is named for the brother, in Greek mythology, of Prometheus, who accepted Pandora (a gift from Zeus) as his wife despite the warning of his brother. As

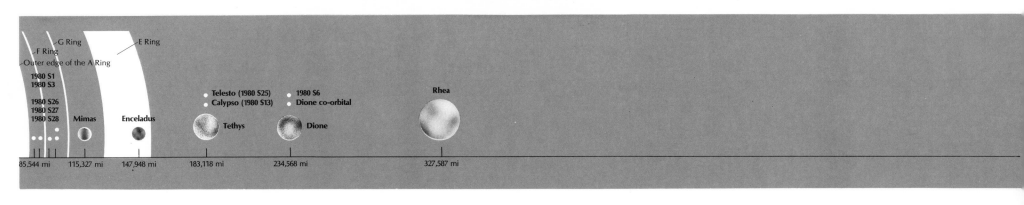

G Ring
F Ring
E Ring
Outer edge of the A Ring
1980 S1
1980 S3
1980 S26
1980 S27
1980 S28
Mimas
Enceladus
Telesto (1980 S25)
Calypso (1980 S13)
Tethys
1980 S6
Dione co-orbital
Dione
Rhea

85,544 mi   115,327 mi   147,948 mi   183,118 mi   234,568 mi   327,587 mi

Prometheus had warned, Pandora opened the infamous box, releasing all the evils within. Janus, the two-faced Roman god of doorways, became the namesake of a moon that was thought to have been identified in 1966 at distance of 105,000 miles from Saturn. The existence of this 'first' Janus was disproven, but the name was reassigned to 1980 S1, which was discovered in a nearby orbit.

Data for Janus (1980S1)

Diameter: 62 × 56 miles (100 × 90 km)
Distance from Saturn: 94,120 miles (151,472 km)
Rotational period (Janusian day): 16.67 Earth hours
Sidereal period (Janusian year): 16.67 Earth hours
Inclination to ecliptic plane: 0°

Data for Epimetheus (1980S3)

Diameter: 56 × 25 miles (90 × 40 km)
Distance from Saturn: 94,089 miles (151,422 km)
Rotational period (Epimethean day): 16.66 Earth hours
Sidereal period (Epimethean year): 16.66 Earth hours
Inclination to ecliptic plane: 0°

Data for Pandora (1980S26)

Diameter: 56 miles (90 km)
Distance from Saturn: 88,048 miles (141,700 km)
Rotational period (Pandoran day): 15.1 Earth hours
Sidereal period (Pandoran year): 15.1 Earth hours
Inclination to ecliptic plane: 0°

Data for Prometheus (1980S27)

Diameter: 136.7 miles (220 km)
Distance from Saturn: 86,589 miles (139,353 km)

Rotational period (Promethean day): 14.67 Earth hours
Sidereal period (Promethean year): 14.67 Earth hours
Inclination to ecliptic plane: 0°

Data for Atlas (1980S28)

Diameter: 24.9 × 12.4 miles (40 × 20 km)
Distance from Saturn: 85,544 miles (137,670 km)
Rotational period (Atlaen day): 14.4 Earth hours
Sidereal period (Atlaen year): 14.4 Earth hours
Inclination to ecliptic plane: Unknown

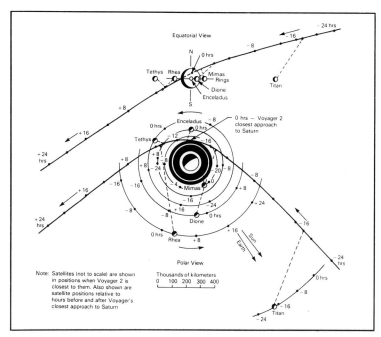

Note: Satellites (not to scale) are shown in positions when Voyager 2 is closest to them. Also shown are satellite positions relative to hours before and after Voyager's closest approach to Saturn.

Thousands of kilometers
0 100 200 300 400

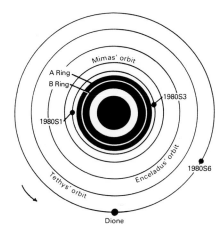

The chart at *left* shows Voyager 2's encounter with Saturn and its moons—and also the angles and distances from which the photos on the preceding and following pages were taken.

Above: After the 1979 ring plane crossing and before the first Voyager Saturn encounter, a total of 12 Saturn moons were known—1980S1 and 1980S3 were co-orbital, and, as one was gaining on the other, it was expected that these two satellites would actually swap orbits as they closed upon one another sometime in 1981 or 1982. With no witnesses to this event, it's currently yet another incidence of 'the tree falling in the forest' syndrome—did they or didn't they, and what was it like?

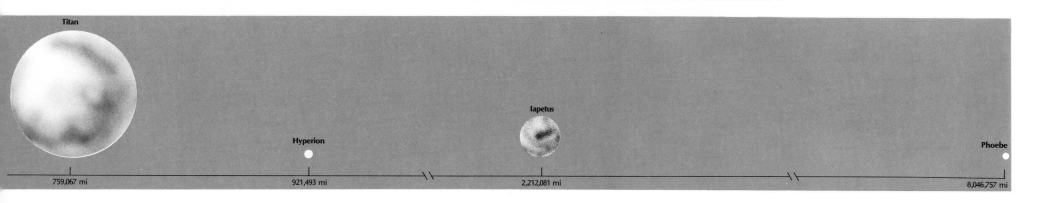

MIMAS

Discovered in 1789 by the German-born but naturalized English astronomer William Herschel (1738–1822), Mimas is scarred by a huge impact crater that bears Herschel's name. This huge crater is centered precisely on the equator and has a diameter one third the diameter of Mimas itself. The walls of the crater Herschel average 16,000 feet, and a huge mountain at the crater's center rises nearly 20,000 feet from its floor. Herschel is truly the standout feature on Mimas, as none of the other impact craters observed on the surface have anywhere near half its diameter. Mimas' surface is also characterized by valleys, or *chasma*, which tend to run in a parallel pattern from southwest to northwest. These uniform valleys are generally 60 miles long, a mile deep and six miles wide. They are thought to be fracture zones which date from the impact which formed Herschel.

Mimas is thought to be composed of 60 percent water ice with roughly 40 percent silicate rock, making it of the 'dirty snowball' class of moons found in much of the outer Solar System. In 1982, Stephen Synnott identified a probable co-orbital moon within the orbit of Mimas.

Above: Mimas, with Herschel turned away.
Facing page: Enceladus showing its 'grooves.'
In the Mercator projections on these pages, zero degrees longitude faces directly toward Saturn, and both Mimas' and Enceladus' leading hemispheres are on the left, and their trailing hemispheres are on the right. Mimas' north polar region and Enceladus' south polar region are still unmapped.

Data for Mimas

Diameter: 242.3 miles (390 km)
Distance from Saturn: 115,326 miles (185,600 km)
Mass: 1.70×10^{19} lb (3.76×10^{19} kg)
Rotational period (Miman day): 22.55 Earth hours
Sidereal period (Miman year): 22.55 Earth hours
Inclination to ecliptic plane: 1.5°
Largest surface feature: Herschel (a huge crater)
80.79 miles diameter (130 km)

Data for Mimas Co-orbital

Diameter: 6.2 miles (10 km)
Distance from Saturn: 115,326 miles (185,600 km)
Rotational period (Miman co-orbital day): 22.55 Earth hours
Sidereal period (Miman co-orbital year): 22.55 Earth hours
Inclination to ecliptic plane: 1.5°

Data for Enceladus

Diameter: 310.7 miles (500 km)
Distance from Saturn: 147,948 miles (238,100 km)
Mass: 3.36×10^{19} lb (7.40×10^{19} kg)
Rotational period (Enceladian day): 1.37 Earth days
Sidereal period (Enceladian year): 1.37 Earth days
Inclination to ecliptic plane: 0°

MERCATOR PROJECTION

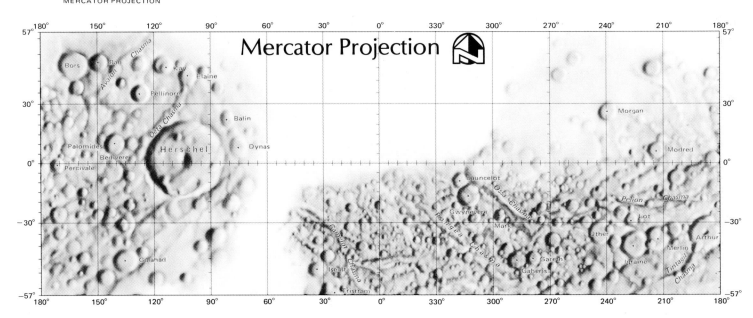

Mercator Projection

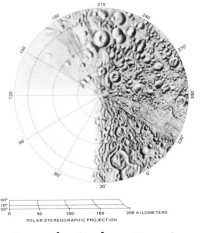

South Polar Region

POLAR STEREOGRAPHIC PROJECTION

ENCELADUS

Discovered by William Herschel (1738–1822) in 1789 at the same time that he identified Mimas, Enceladus is the most geologically active of Saturn's moons. It is named for the giant who rebelled against the gods of Greek mythology, and who was subsequently struck down and buried on Mount Etna.

Like Mimas, Enceladus is composed mostly of water with the remaining roughly 40 percent of its material being silicate rock. Enceladus, however, has a much more complex surface which is divided between vast and ancient fields of impact craters, large smooth plains and complex mountain ranges. The latter are typical of the type of fracture zones that characterize the surfaces of Jupiter's moon Ganymede or Uranus's moon Miranda. These features, including ridges and valleys, were possibly formed by the same sort of pressure between separate surface plates that is responsible for the silicate rock mountain ranges on Earth and the ice mountain ranges on Ganymede.

The smooth plains on Enceladus are further evidence of fracture zones because they were possibly formed by liquid water welling up from the interior and spilling out through fissures and faults, forming lakes in lowland areas. These lakes covered older fields of impact craters and when the water froze, the lakes became the smooth and newer plains which we see today.

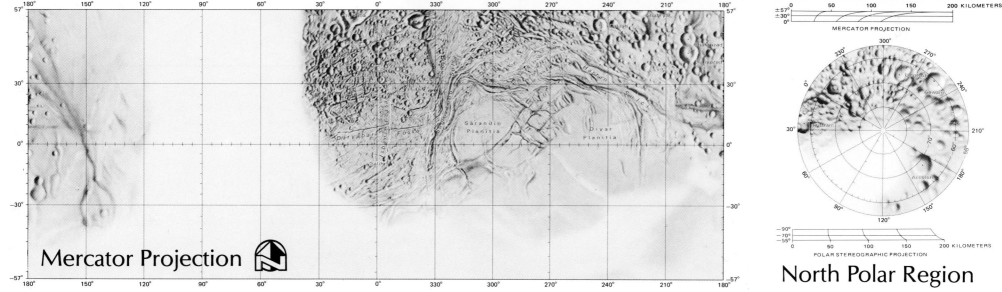

Mercator Projection

North Polar Region

Above: Voyager 2 obtained this image of Tethys from a distance of 368,000 miles. Tethys shows two distinct types of terrain—bright, densely cratered regions; and relatively dark, lightly cratered planes. The densely cratered terrain is believed to be part of the ancient crust of the satellite; the lightly cratered planes are thought to have been formed later by internal processes. Also seen here is a canyon trough that runs parallel to the terminator (the day-night boundary, seen at the right); this is an extension of the huge canyon system Ithaca Chasma, which extends nearly two-thirds the distance around Tethys.

At right: Saturn and two of its moons, Tethys (above) and Dione, were photographed by Voyager 1 on 3 November 1980. The shadows of Saturn's three bright rings and Tethys are cast onto the cloud tops.

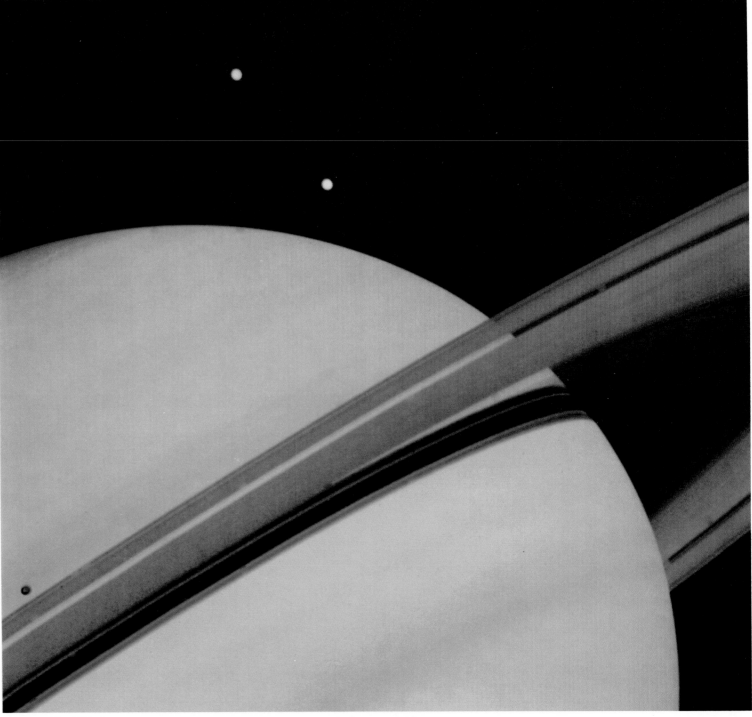

TETHYS

Discovered in 1684 by Giovanni Domenico Cassini (1625–1712), Tethys is named for the Greek sea goddess who was both wife and sister of Oceanus. Like the smaller Mimas and Enceladus, Tethys is a 'dirty snowball' composed mostly of water ice, with silicate rock as a secondary component. Most of its surface is marred by impact craters, but they appear somewhat softened—as though the surface had warmed slightly at some point since the formation of the craters, and partial melting had taken place.

The most notable feature on Tethys is the mysterious Ithaca Chasma, an enormous *rift canyon* that runs from near the north pole all the way to the south pole. With an average width of 60 miles and an average depth of three miles, Ithaca Chasma dwarfs the Earth's Grand Canyon in both scalar and absolute terms. In the scale of the Earth, the equivalent of Ithaca Chasma would be like having a 40 mile deep trench as wide as the state of Colorado extending from Nome, Alaska to the southern tip of Argentina. Of uncertain origin, this huge canyon was probably formed when Tethys cooled after it was first formed.

Tethys also has the distinction of being the 'parent,' both figuratively and *perhaps* literally, of a family of as many as *five*

In the Mercator projection at *below left,* Tethys' leading hemisphere (in terms of its orbital motion) is on the left and its trailing hemisphere is on the right; zero degrees longitude faces directly toward Saturn.

At left: Tethys shows the complement to the view on the facing page.

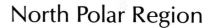

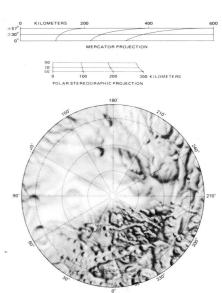

North Polar Region

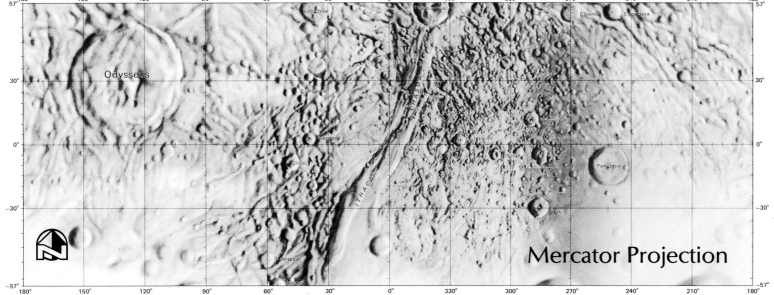

Mercator Projection

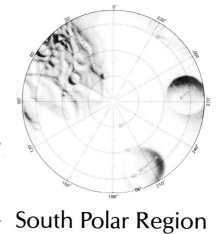

South Polar Region

In the Mercator projection *at below right*—as in similar projections in this chapter—Dione's leading hemisphere is on the left and its trailing hemisphere is on the right, and zero degrees longitude directly faces Saturn. Compare the maps (*below*) of Dione's polar region with those of Mimas and Enceladus on pages 134-135.

Data for Tethys

Diameter: 652.4 miles (1050 km)
Distance from Saturn: 182,714 miles (294,700 km)
Mass: 2.85×10^{21} lb (6.32×10^{21} kg)
Rotational period (Tethian day): 1.89 Earth days (45.3 hours)
Sidereal period (Tethian year): 1.89 Earth days
Inclination to ecliptic plane: 1.1°
Largest surface feature: Ithaca Chasma 1550 miles long

Data for Telesto

Diameter: 9.3 miles (15 km)*
Distance from Saturn: 183,118 miles (294,700 km)
Rotational period (Telestan day): 1.9 Earth days
Sidereal period (Telestan year): 1.9 Earth days
Inclination to ecliptic plane: 0°

Data for Calypso

Diameter: 9.3 miles (15 km)*
Distance from Saturn: 217,479.91 miles (350,000 km)
Rotational period (Calypsan day): 2.44 Earth days
Sidereal period (Calypsan year): 2.44 Earth days
Inclination to ecliptic plane: Unknown
*Some data suggests a diameter of 15.5 miles.

co-orbital moons. The larger of these are Telesto and Calypso (formerly 1980S25 and 1980S13). Calypso is named for the sea nymph of Greek mythology who delayed Odysseus and turned his men into pigs on her island. Calypso is derived from Greek *Kalupso,* meaning 'she who conceals'—an apt reference for a member of the group of co-orbitals which lurk in the shadow of Tethys.

The physical nature of the Tethys co-orbital family is unknown, but it would probably not be out of line to suggest that they are 'dirty snowballs' with surfaces that have been battered by meteorites.

DIONE

Discovered by Giovanni Cassini (1625–1712) in 1684, Dione is named for the mother of the Greek goddess Aphrodite. Like the moons closer to Saturn, Dione is composed of silicate rock and water ice. However, while the inner moons have a predominance of ice, Dione is at least half rock.

Dione's surface is darker than any of Saturn's other ice/rock moons, indicating that there are large regions of exposed rock. Dione's surface is characterized by impact craters common to

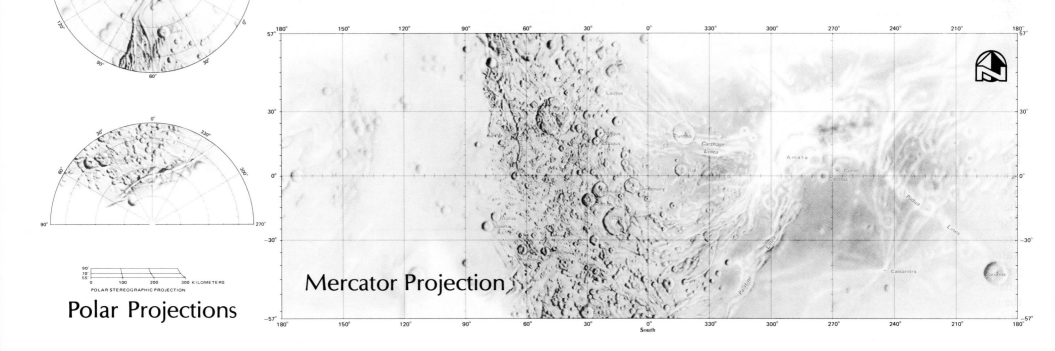

Polar Projections

Mercator Projection

Data for Dione

Diameter: 695.9 miles (1120 km)
Distance from Saturn: 234,567 miles (377,500 km)
Mass: 4.77×10^{20} lb (1.05×10^{21} kg)
Rotational period (Dionian day): 2.74 Earth days (65.7 hours)
Sidereal period (Dionian year): 2.74 Earth days
Inclination to ecliptic plane: 0°
Largest surface feature: Amata 149.13 miles in diameter

Data for Dione Co-orbital I

Diameter: 9.3 miles (15 km)
Distance from Saturn: 234,878 miles (378,000 km)
Rotational period (Dionian co-orbital day): Unknown
Sidereal period (Dionian co-orbital year): Unknown
Inclination to ecliptic plane: .3°

Data for Dione Co-orbital II

Diameter: Unknown
Distance from Saturn: 291,982 miles (469,900 km)
Rotational period (Dionian co-orbital day): Unknown
Sidereal period (Dionian co-orbital year): Unknown
Inclination to ecliptic plane: Unknown

Data for 1980S6

Diameter: $22 \times 20 \times 19$ miles ($35 \times 32 \times 31$ km)
Distance from Saturn: 234,915 miles (378,060 km)
Rotational period (1980S6 day): 2.7 Earth days (65.59 hours)
Sidereal period (1980S6 year): 2.7 Earth days
Inclination to ecliptic plane: Unknown

both icy and rocky surface areas. The craters are generally smaller than 25 miles across, but Amata, the largest, measures nearly 150 miles in diameter. Amata is coincidentally located at the center of an eastern hemisphere pattern of unusual light colored streaks. While the streaks on Enceladus and Tethys are very sharply defined as though cut with a knife, Dione's streaks are wispy—as though they were painted with an air brush. These streaks are probably cracks or fissures through which liquid water seeped and refroze over time. Amata appears to have been partially inundated by this seepage and it has been suggested that whatever impact created Amata may have also played a role in the formation of the wispy streaks because many of them seem to radiate from the crater.

Dione's family of co-orbitals includes 1980 S6 which is sometimes called Dione B. The largest of the Lagrangians, 1980 S6 is the same size as Phoebe, the outermost of Saturn's moons. S6 was discovered in 1980 by Lacques and Lecacheux in France and photographed by the Voyager spacecraft just a few months later. It was an interesting coincidence that Saturn's rings were oriented toward the Earth in such a way that it was possible for groundbased observers to discover 1980 S6 in the same year that the Voyagers made their own spectacular discoveries.

Stephen Synnott has identified a second Dione co-orbital that might actually be a 1980 S6 co-orbital, as the center lines of the orbits of these bodies are offset from Dione's path by about 300 miles, or exactly Dione's radius.

A third possible member, of either the Dione family or the Rhea family, is a body described by Synnott which is located about 57,000 miles beyond the orbital path of Dione and 1980 S6, and 35,300 miles inward from Rhea's orbit.

Above: The craters Aeneas, Romulus, Remus and Dido are clearly visible in this photo of Dione's icy visage taken by Voyager at a distance of 100,440 miles. Many impact craters are show in this color mosaic. The largest crater is less than 62 miles in diameter and shows a well-developed central peak. Sinuous fault valleys break the moon's icy crust.

Below: In this NASA painting by Ron Miller, Saturn is represented as it might appear when viewed from the surface of Rhea. Rhea is an icy wasteland and is completely without an atmosphere; therefore, Rhea's mother planet may well be seen very clearly from the moon's surface.

At right: Color enhancement was used to bring out subtle variations in color and brightness in this Voyager 1 view of Rhea. The white streaks that cross Rhea's surface are probably ice that has just recently been ejected from beneath the moon's surface—this 'repaving' is part of the reason that Rhea is so bright.

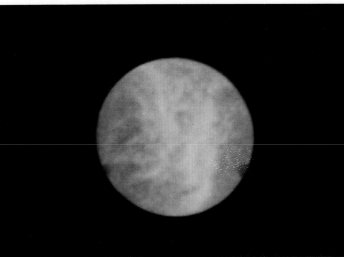

RHEA

The second largest of Saturn's moons, Rhea was discovered by Giovanni Cassini (1625–1712) in 1672 and was named for the wife of Kronos who, according to Greek mythology, ruled the universe until dethroned by his son Zeus. In Roman mythology as well as in astronomical nomenclature, Rhea is identified with Saturn because Saturn is the father of Jupiter, the Roman equivalent of Zeus.

Rhea the moon is composed of an equal mixture of water ice

Mercator Projection

and silicate rock, with a possible core of denser material or perhaps solid rock. The moon's leading hemisphere is solid ice with intermittent patches of frost, while the darker, trailing eastern hemisphere shows signs of the same wispy surface detail that was observed more minutely by the Voyager spacecraft on Dione. The icy western hemisphere is characterized exclusively by thousands of impact craters, while there seem to be fewer craters in the eastern hemisphere and in the equitorial regions of both hemispheres. The largest crater, Izanagi, located deep in the southern hemisphere near the *prime meridian,* has a diameter of nearly 140 miles.

Data for Rhea

Diameter: 950.7 miles (1530 km)
Distance from Saturn: 327,586 miles (527,200 km)
Mass: 1.0×10^{21} lb (2.28×10^{21} kg)
Rotational period (Rhean day): 4.52 Earth days (108 hours)
Sidereal period (Rhean year): 4.52 Earth days
Inclination to ecliptic plane: .3°

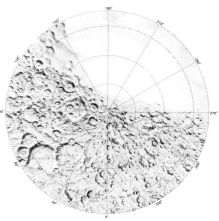

North Polar Region

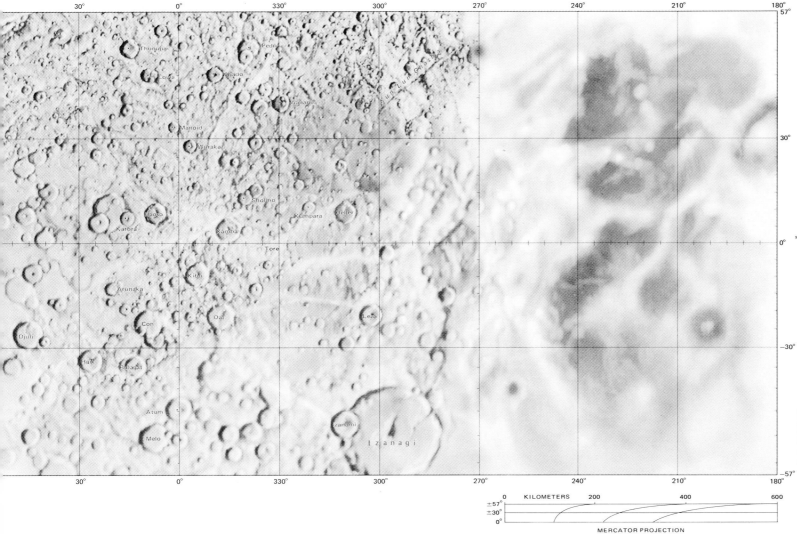

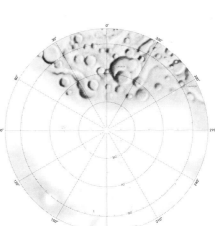

South Polar Region

At left: Rhea's leading hemisphere is on the left of this Mercator projection, its trailing hemisphere is on the right, and zero degrees longitude directly faces Saturn. Large parts of Rhea's southern hemisphere remain uncharted.

TITAN

Above: Saturn's shrouded moon, the gargantuan Titan. We have no map of Titan because its surface is hidden beneath a thick hydrocarbon atmosphere. The charting of Titan's oceans and land masses will have to await the type of radar mapping that allowed us to peer beneath the cloud cover of Venus. The southern hemisphere appears lighter than the northern, and a well defined darker band can be seen near the equator—both intriguing shadows of the wonders that remain to be discovered on Saturn's largest moon.

Facing page: Layers of haze cover Titan in this false color image taken by Voyager 1 on 12 November 1980 at a range of 13,700 miles. The upper level of the thick aerosol above the satellite appears orange.

Larger than either of the planets Mercury and Pluto, Titan is the second largest moon in the Solar System (after Ganymede). Once thought to be the largest, it was named for the family of pre-Olympian Greek gods whose appellation implies colossal size. Titan was discovered by the Dutch astronomer Christian Huygens (1629–1695) in 1655, making it one of the first moons to be discovered in the Solar System. Though no longer on the throne as the Solar System's largest moon, it certainly must rank as one of the most spectacular.

Titan is the only known moon with a fully developed atmosphere that consists of more than simply trace gases. Titan has, in fact, a denser atmosphere and cloud cover than either Earth or Mars. This cloud cover, nearly as opaque as that which shrouds Venus, has prevented the sort of surface mapping that has been possible with the other major moons of the outer Solar System, but its presence has only served to make Titan all that much more intriguing. (Early astronomers mistook this dense atmosphere for Titan's actual surface, and it was through this mistake that Titan was once considered to be the Solar System's largest moon.)

Titan's atmosphere is extremely rich in nitrogen, the same element that makes up the greatest part of the Earth's atmosphere. Other major components of Titan's atmosphere are hydrocarbon gases such as acetylene, ethane and propane, with methane being the most common of the hydrocarbons. While these gases are also to be found in Saturn's own atmosphere, Titan's atmosphere contains four times the ppm concentration of ethane and 150 times the concentration of acetylene. Titan's atmosphere probably includes broken methane clouds at an altitude of about 25 miles, with the dense, smoggy hydrocarbon haze stretching up to an altitude of nearly 200 miles—where ultraviolet radiation from the Sun converts methane to acetylene or ethane.

Why Titan was able to develop an atmosphere while Ganymede and Callisto (bodies of similar size) did not, is a matter of conjecture. It has been theorized that all the Solar System's largest moons had similar chemistry in the beginning, but Titan evolved in a colder part of the Solar System—farther away from the Sun, and Jupiter when it almost became a primordial star. Thus the hydrocarbon gases were able to exist as solids on Titan while the gases of Jupiter's Galilean moons dissipated into space, leaving only water and rock. Saturn's other moons were never large enough to have sufficient gravity to hold an atmosphere.

The view from Titan's surface is one of an exciting but inhospitable world. Covered by the opaque haze, the sky would appear like a smoggy sunset on Earth or like a view from the surface of Venus. The atmospheric pressure on Titan's surface, while 1.6 times that of Earth, is however a good deal less than that on Venus. Titan's surface temperature of nearly −300 degrees Fahrenheit would permit methane to exist not only as a gas, but also as a liquid or solid much in the same way that water does on Earth. A picture is thus painted of a cold orange-tinted land where methane rain or snow falls from the methane clouds and where methane rivers may flow into methane oceans dotted with methane icebergs. There is evidence of a 30 Earth-year seasonal cycle which *may* permit the development of methane ice caps that expand and recede like the water ice caps on Earth (and the water/carbon dioxide ice caps on Mars). Water ice is also present on Titan, beneath the methane surface features, and possibly extends up into the atmosphere in the form of ice mountains. Titan's mantle is, in turn, largely composed of water ice that gives way to a rocky core perhaps 600 miles beneath the surface. The absence of a magnetic field indicates that Titan has no significant amount of ferrous metallic minerals in its core.

The presence of nitrogen, a hydrocarbon atmosphere and water indicate that Titan's surface is very much like that of the Earth four billion years ago, before life evolved on the latter body. It has been suggested that this similarity to the prebiotic 'soup' that covered the Earth in those bygone days could presage a similar chain of events on Titan.

Data for Titan
Diameter: 3200 miles (5150 km)
Distance from Saturn: 759,067 miles (1,221,600 km)
Mass: 6.1818×10^{23} lb (1.36×10^{23} kg)
Rotational period (Titanian day): 15.9 Earth days (383 hours)
Sidereal period (Titanian year): 15.9 Earth days
Inclination to ecliptic plane: .3°
Mean surface temperature: −288° F
Major atmospheric components: Nitrogen (94%)
Other atmospheric components: Methane, Helium, Ethane, Diacetylene, Hydrogen cyanide, Cyanogen, Methylacetylene, Cyanoacetylene, Acetylene, Propane, Carbon dioxide, Carbon monoxide

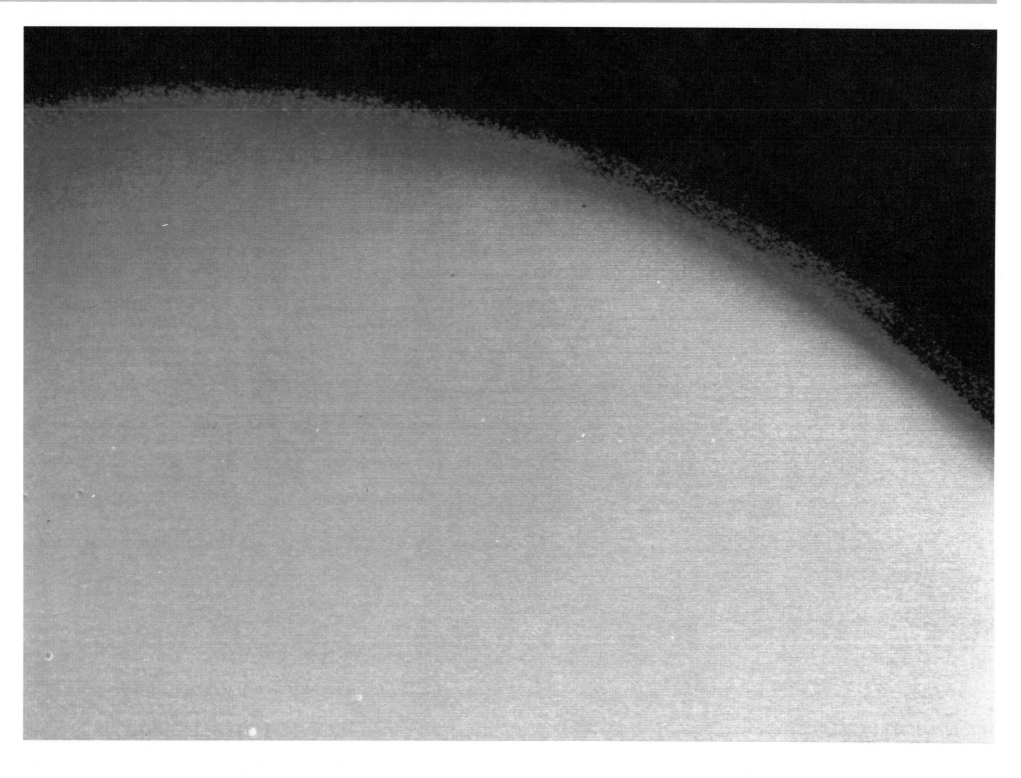

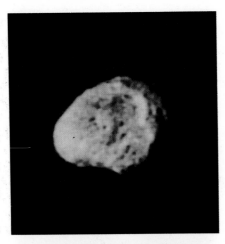

HYPERION

A dark, rocky moon, Hyperion is less completely ice-covered than its larger brothers closer to Saturn, although water-ice is prevalent over much of its crater-pocked surface. Though it has several craters with diameters in excess of 30 miles, Hyperion's most distinguishing feature is its irregular shape. Elliptical shapes are common among smaller moons throughout the Solar System, but it is uncommon to find one of Hyperion's size that is not more perfectly spherical. The mystery is deepened by the fact that the long axis is not oriented directly toward Saturn, which tends to indicate that Hyperion may have been the victim of a collision with another body—such as a large meteorite—at some time in its relatively recent geologic history.

In Greek mythology, Hyperion's namesake was one of the Titans, a son of Uranus and Gaea, and father of Helios, Selene, and Eos (the Sun, the Moon and the dawn). Hyperion was often referred to by the Greek poets as the Sun god and as such was often identified with Apollo.

Data for Hyperion
Diameter: 249 × 155 × 149 miles (400 × 250 × 240 km)
Distance from Saturn: 921,493 miles (1,483,000 km)
Mass: 5×10^{19} lb (1.1×10^{20} kg)
Rotational period (Hyperionian day): 21.28 Earth days
(510.7 hours)
Sidereal period (Hyperionian year): 21.28 Earth days
Inclination to ecliptic plane: .6°
Largest surface feature: Scarp system 186.4 miles (300 km)
in length

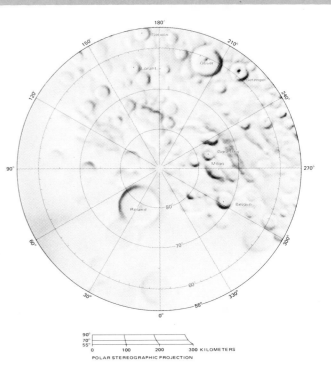

PHOEBE

The outermost of Saturn's known moons, Phoebe was discovered in 1898 by American astronomer William Henry Pickering (1858-1938). Phoebe has several features that distinguish it from all the other moons of Saturn: non-sychronous rotation, a retrograde orbit—and the plane of that orbit, which is tilted 150 degrees from Saturn's equatorial plane. This is *10 times* the inclination exhibited by the orbit of Iapetus, and all the other moons—

from tiny Atlas to gigantic Titan—orbit within *two* degrees of Saturn's equatorial plane.

Because of these unusual characteristics and because of Phoebe's irregular shape, it has been suggested that this moon may be either an asteroid or a comet nucleus that has been trapped into orbit by Saturn's gravity. The fact that Phoebe seems to be more rocky than icy would tend to mitigate against the comet nucleus theory.

In Greek mythology, Phoebe was a title given to Artemis in her role as goddess of the Moon. It is a parallel to Phoebus, the title given to Apollo in his character of Sun god.

Data for Phoebe
Diameter: 136.9 miles (220 km)
Distance from Saturn: 8,046,756 miles (12,950,000 km)
Rotational period (Phoeben day): 9 Earth hours
Sidereal period (Phoeben year): 550.3 Earth days
Inclination to ecliptic plane: 150°

North Polar Region

IAPETUS

Discovered in 1671 by Giovanni Cassini (1625-1712), Iapetus is the third largest of Saturn's moons in terms of size. In those terms it is, in fact, a near twin of Rhea, but is a good deal less dense. As such, it probably has a higher proportion of ice in relation to rock in its composition. The most intriguing thing about Iapetus is that its entire leading hemisphere is as black as asphalt while its trailing hemisphere contains a more familiar crater-pocked ice and rock landscape. The nature and source of this vast dark area is unknown, but some crater floors on the trailing hemisphere appear to be equally dark and this tends to suggest that the black material is extruded from beneath an otherwise icy surface. The fact that this dark area is apparently unblemished by icy, light colored meteorite craters indicates that the material—whatever it is—probably is renewed regularly. One is reminded of the molten sulfur flows that are constantly resurfacing Jupiter's moon Io, or the lava flows in the Earth's Hawaiian islands.

Like his brother, the mythological Hyperion, the Iapetus of Greek mythology was a Titan. Son of Uranus and Gaea, he was the father of Atlas, Epimetheus, Menoetius and Prometheus. According to Greek legend, Iapetus was imprisoned by Zeus in Tartarus after the rebellion of the Titans against the gods.

Data for Iapetus
Diameter: 905.2 miles (1460 km)
Distance from Saturn: 2,212,081 miles (3,560,000 km)
Mass: 8.77×10^{20} lb (1.93×10^{21} kg)
Rotational period (Iapetan day): 79.33 Earth days
(1904 hours)
Sidereal period (Iapetan year): 79.33 Earth days
Inclination to ecliptic plane: 14.7°
Largest surface feature: Dark ring extending beyond dark hemisphere approximately 249 miles (400 km) in diameter

Facing page: The Saturnian moon Hyperion (*top*) here shows its irregular boulder-like shape. The moon is 235 miles across. The north pole of Iapetus is near the large central-peak crater seen partly in the shadow at the top of the *middle photo.* Note the *Cassini Regio,* or region of very dark material (lower and right hand parts of the photo) that apparently covers the ice crust of the satellite primarily at its leading hemisphere. *Bottom photo:* Saturn's outermost known satellite, Phoebe, is possibly a captured asteroid.

The polar view and Mercator projection *at lower left* both portray Iapetus, third largest of Saturn's moons. The moon's leading hemisphere is at the left of the map, its trailing hemisphere is at the right, and zero degrees longitude directly faces Saturn.

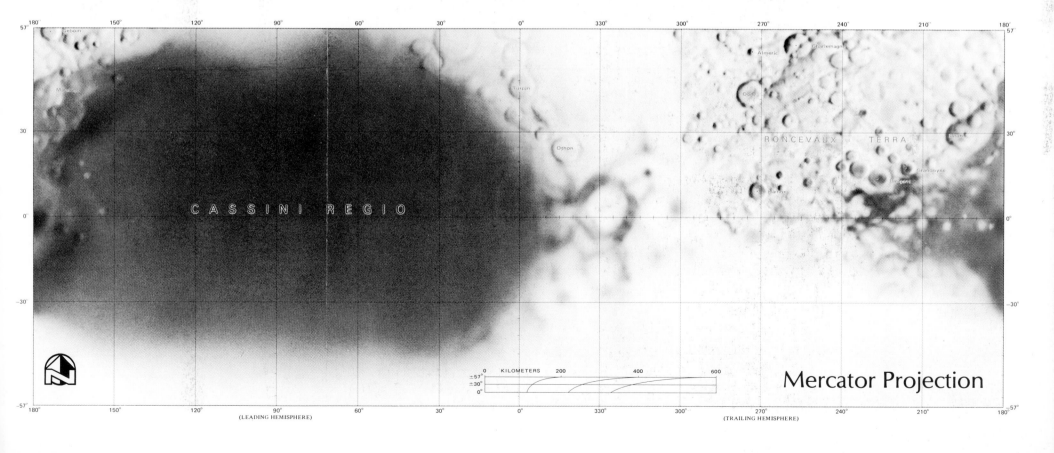

Mercator Projection

URANUS

The first outer Solar System planet to be correctly identified as such in historical times, Uranus was identified in 1781 by the German-born English astronomer William Herschel (1738–1822) while he was working at Bath. The planet is named for the earliest supreme god of Greek mythology. The personification of the sky, mythical Uranus was both son and consort to the goddess Gaea and father of all the Cyclopes and Titans.

Uranus has an axial inclination of 98 degrees. Discovered in 1829, this is a phenomenon that is unique in the Solar System. With such an axial inclination Uranus is seen as rotating 'on its side' at a near right angle to the inclination of the Earth or Sun. The poles of Uranus rather than its equatorial regions are pointing alternately at the Sun.

Uranus is a gaseous planet like Jupiter, Saturn and Neptune, with a distinct blue-green appearance probably due to a concentration of methane in its upper *atmosphere*. In terms of size, it is smaller than Jupiter and Saturn while being very close to the size of Neptune. Its solid *core* is composed of metals and silicate rock with a diameter of roughly 270,000 miles. Its core is in turn covered by an icy *mantle* of methane ammonia and water ice 6000 miles deep.

As with the others gaseous planets, the predominant elements in the Uranian atmosphere are hydrogen and helium, although the Voyager 2 observations in 1986 indicated that the atmosphere was only 15 percent helium versus 40 percent as originally postulated. Other atmospheric constituents include methane, acetylene and other hydrocarbons.

The clouds that form in this atmosphere are moved by prevailing winds that blow in the same direction as the planet rotates, just as they do on Jupiter, Saturn and Earth. The planet's lowest atmospheric temperature (– 366 degrees Fahrenheit) is recorded at the boundary between the *troposphere* and *stratosphere*. Surprisingly, both the poles show similar temperatures whether or not they are sunlit. The coldest latitudes seem to be those between 15 and 40 degrees. In the upper atmosphere, temperatures increase to – 190 degrees Fahrenheit, while in the planet's interior, temperatures are extremely hot—as are the interiors of Jupiter and Saturn.

Prior to the flyby of Voyager 2 in January 1986, Uranus was thought not to have a magnetic field, but this assumption proved false. The magnetic field of Uranus is tilted at a 60 degree angle to the planet's rotational axis (compared to 12 degrees on Earth). The magnetic field has roughly the same intensity as the Earth's, but whereas the Earth's magnetic field is generated by a molten metallic core, the one surrounding Uranus seems to be generated by the electrically conductive, super-pressurized ocean of ammonia and water that exists beneath the atmosphere.

Uranus, like Jupiter and Saturn, has a system of *rings,* of which the first nine were discovered by Earth-based observers in 1977. In 1986, Voyager 2 observed these in detail, and identified two more. This ring system is much more complex than that of Jupiter, but less so than Saturn's spectacular system. The system around Uranus seems to be relatively young and probably did not form at the same time as the planet. The particles that make up the rings may be the remnants of a moon that was either broken by a high velocity impact or torn apart by the gravitational effects of Uranus.

The widest ring known before Voyager 2 was the outermost ring, Epsilon—an *irregular* ring measuring 14 to 60 miles across. In 1986, Voyager's cameras helped identify a new innermost ring, designated 1986 U2R, that is 1550 miles wide. The narrowest complete rings are less than a mile wide, while faint possibly

Facing page: This view of Uranus was recorded by Voyager 2 on 25 January 1986, as the spacecraft left the planet behind and set forth on its cruise to Neptune. Voyager was about 600,000 miles from Uranus when it acquired this wide angle view. The thin crescent of Uranus is seen here at an angle of 153 degrees between the spacecraft, the planet and the Sun.

Even at this extreme angle, Uranus retains the pale blue-green color seen by ground-based astronomers and recorded by Voyager during its historic encounter. This color results from the presence of methane in the Uranian atmosphere; the gas absorbs red wavelengths of light, leaving the predominant hue seen here. The tendency for the crescent to become white at the extreme edge is caused by the presence of a high-altitude haze.

incomplete rings have been identified which are only 160 feet across. The rings are composed of large blocks of ice with small dust particles scattered throughout the system. The outer edge of the system, the outer edge of the Epsilon Ring, is sharply defined and is 15,800 miles from the Uranus cloud tops. At this point the Epsilon Ring is just 500 feet thick, and is surprisingly devoid of fragments with diameters below one foot.

THE URANIAN MOONS

Prior to the observations by Voyager 2, Uranus was known to have just five moons. Photographs returned by the spacecraft increased the number of known moons to 15, with all 10 of the newly-discovered moons located *within* the orbital paths of the original five. One of the new moons, 1985 U1, was discovered by Voyager's cameras in late 1985 and the rest were discovered in the photos taken during the January 1986 Voyager flyby of the Uranian system. With the exception of 1985 U1 and 1986 U7—the largest and smallest of the 'Voyager' moons—all of the newly discovered members of the group are very uniform in size, with diameters ranging between 31 and 37 miles.

The innermost of the moons are 1986 U7, located between the Delta Ring and the Epsilon Ring, and 1986 U8, on the opposite side of the Epsilon Ring. Thus straddling the Epsilon Ring, these two small bodies probably act like the shepherd moons of Saturn, controlling and defining the position and shape of the ring.

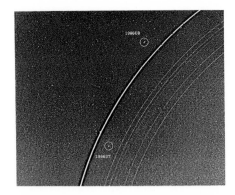

Above: Voyager 2 took this photo of Uranus' Epsilon Ring with its two shepherd moons, 1986U7 (lower moon) and 1986U8 (upper moon)—these also appear in the chart of the Uranian ring system and the Uranian moons which is shown *below.* Depending on the criteria used by various astronomers, distances from the rings and moons to the planet are sometimes given as measured to the tops of the planet's considerably thick outer cloud layer, and sometimes are given as measured to the very dense gas atmosphere which lies under the cloud layer.

Data for Uranus

Diameter: 32,116 miles (51,800 km)
Distance from Sun: 1,862,480,000 miles
$\qquad$ (3,004,000,000 km) at aphelion
$\qquad$ 1,695,700,000 miles
$\qquad$ (2,735,000,000 km) at perihelion
Mass: 3.953×10^{25} lb (8.698×10^{25} kg)
Rotational period (Uranian day): 17.3 Earth hours
$\qquad$ (retrograde)
Sidereal period (Uranian year): 84 Earth years
Eccentricity: 0.047
Inclination of rotational axis: 82.1°
Inclination to ecliptic plane (Earth = 0): .77
Albedo (100% reflection of light = 1): .34 to .5
Mean temperature: −350° F
Maximum temperature: −190° F
Minimum temperature: −366° F
Largest feature: Not visible
Major atmospheric components: Hydrogen (85%)
$\qquad$ Helium (15%)
Other atmospheric components: Ammonia, Sulfur,
$\qquad$ Methane, Acetylene,
$\qquad$ other hydrocarbons

Data for 1985U1

Diameter: 105.6 miles (170 km)
Distance from Uranus: 53,437 miles (86,000 km)
Rotational period: Unknown
Sidereal period: Unknown
Inclination to ecliptic plane: Unknown

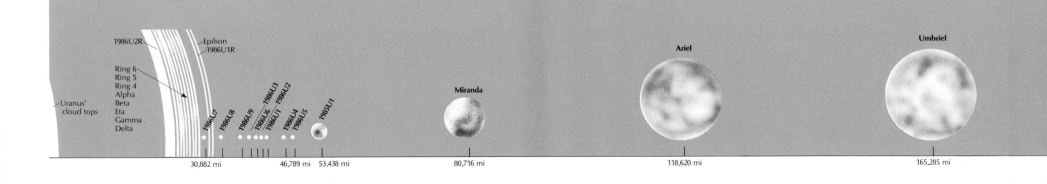

30,882 mi   46,789 mi 53,438 mi   80,716 mi   118,620 mi   165,285 mi

Data for 1986U1

Diameter: 49.71 miles (80 km)
Distance from Uranus: 41,072 miles (66,100 km)
Rotational period: Unknown
Sidereal period: Unknown
Inclination to ecliptic plane: Unknown

Data for 1986U2

Diameter: 49.71 miles (80 km)
Distance from Uranus: 40,140 miles (64,600 km)
Rotational period: Unknown
Sidereal period: Unknown
Inclination to ecliptic plane: Unknown

Data for 1986U3

Diameter: 37.28 miles (60 km)
Distance from Uranus: 38,400 miles (61,800 km)
Rotational period: Unknown
Sidereal period: Unknown
Inclination to ecliptic plane: Unknown

Data for 1986U4

Diameter: 37.28 miles (60 km)
Distance from Uranus: 43,433 miles (69,900 km)
Rotational period: Unknown
Sidereal period: Unknown
Inclination to ecliptic plane: Unknown

Data for 1986U5

Diameter: 37.28 miles (60 km)
Distance from Uranus: 46,789 miles (75,300 km)

Rotational period: Unknown
Sidereal period: Unknown
Inclination to ecliptic plane: Unknown

Data for 1986U6

Diameter: 37.28 miles (60 km)
Distance from Uranus: 38,959 miles (62,700 km)
Rotational period: Unknown
Sidereal period: Unknown
Inclination to ecliptic plane: Unknown

Data for 1986U7

Diameter: 24.86 miles (40 km)
Distance from Uranus: 30,882.15 miles (49,700 km)
Rotational period: Unknown
Sidereal period: Unknown
Inclination to ecliptic plane: Unknown

Data for 1986U8

Diameter: 31.07 miles (50 km)
Distance from Uranus: 33,429 miles (53,800 km)
Rotational period: Unknown
Sidereal period: Unknown
Inclination to ecliptic plane: Unknown

Data for 1986U9

Diameter: 31.07 miles (50 km)
Distance from Uranus: 36,785 miles (59,200 km)
Rotational period: Unknown
Sidereal period: Unknown
Inclination to ecliptic plane: Unknown

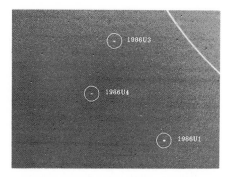

Three of the Uranian moons which were discovered by Voyager 2 appear in the photo *above*. From the top, they are 1986U3, 1986U4 and 1986U1. Uranus' Epsilon Ring is at the upper right.

The Uranian Ring System

Ring Composition	Orbital Distance (Planet center to inner edge of ring in miles)	Ring width (in miles)
1986U2R	1900	24,000
Ring 6	.62 to 1.86	25,978
Ring 5	1.24 to 1.86	26,226
Ring 4	1.24	26,412
Alpha	4.96 to 6.82	27,776
Beta	4.34 to 6.82	28,334
Eta	1.24	29,264
Gamma	.62 to 2.48	29,574
Delta	1.86 to 5.58	29,946
1986U1R	.62 to 1.24	31,025
Epsilon	13.64 to 57.66	31,744

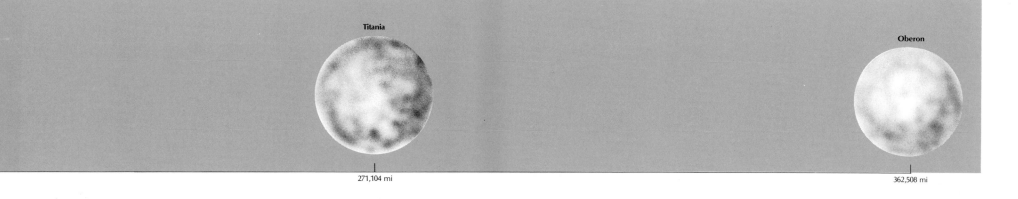

Titania
271,104 mi

Oberon
362,508 mi

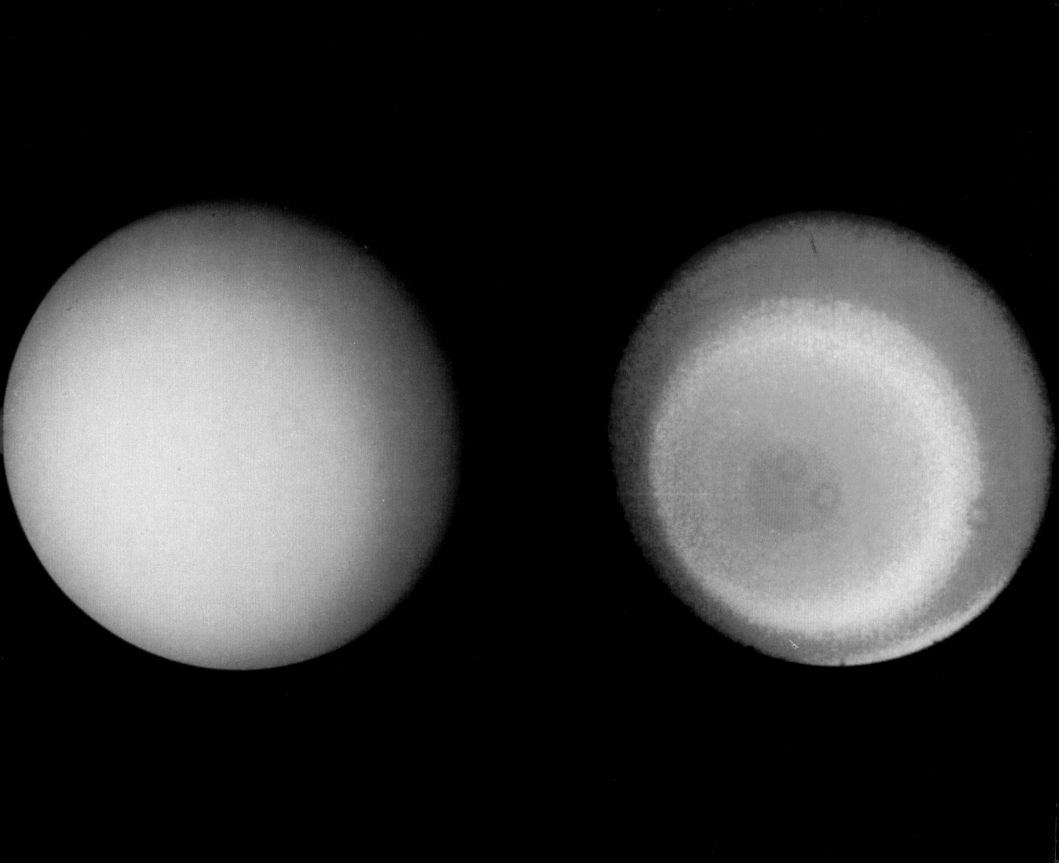

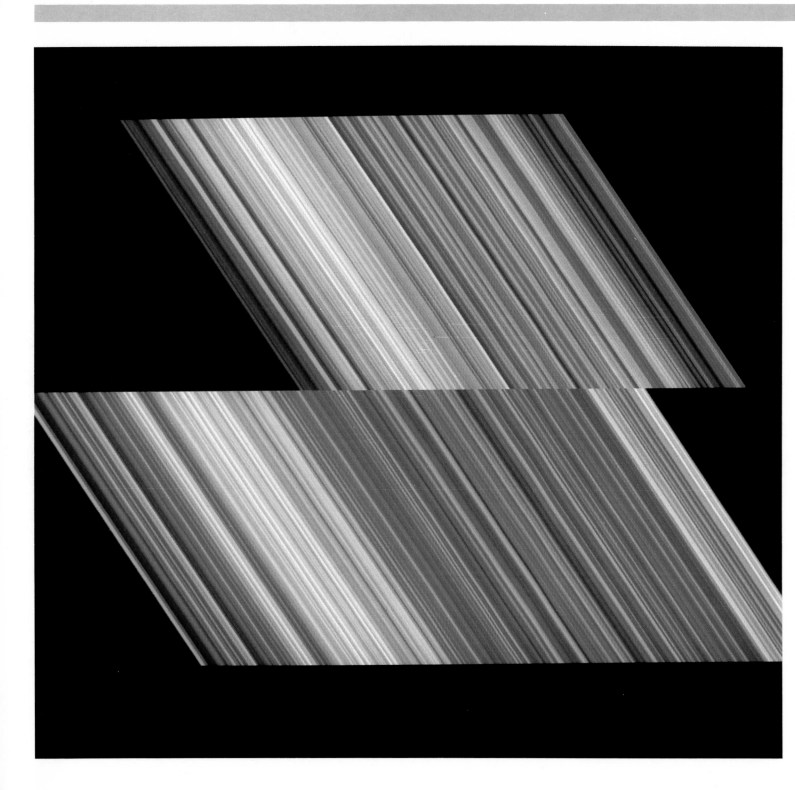

Facing page: These two pictures of Uranus—one in true color (left) and the other in false color—were compiled from images returned on 17 January 1986 by Voyager 2 spacecraft from a distance of 5.7 million miles. The picture on the left has been processed to show Uranus as human eyes would see it.

Note the blue green color of Uranus' deep, cold and remarkably clear atmosphere. The darker shadings at the upper right of the disk correspond to the day-night boundary on the planet. Beyond this boundary lies the northern hemisphere of Uranus that remains in total darkness as the planet rotates.

The picture on the right uses false colors and contrast enhancement to bring out subtle details in the polar region of Uranus. In this picture, Uranus reveals a dark polar hood surrounded by a series of progressively lighter concentric bands. One possible explanation is that a brownish haze or smog, concentrated over the pole, is arranged into bands by zonal motions of the upper atmosphere.

The occasional doughnut shapes are shadows cast by dust in the camera's optics.

At left: Two slices of Uranus' Epsilon Ring are shown in this image created from the photopolarimeter aboard the Voyager 2 spacecraft. The measurements show the structure and amount of material in the rings. This false-color image enhances the visibility of small features, but doesn't show that the rings are really very dark. The disparity in the size of the two slices illustrates the accordion-like nature of the ring, which seems to contract and expand as it orbits the planet.

The false-color photo *above* of Uranus's rings is from images taken by Voyager 2 on 21 January 1986, from a distance of 2.59 million miles. All nine known rings are visible here. The somewhat fainter, pastel lines seen between them are contributed by the computer enhancement.

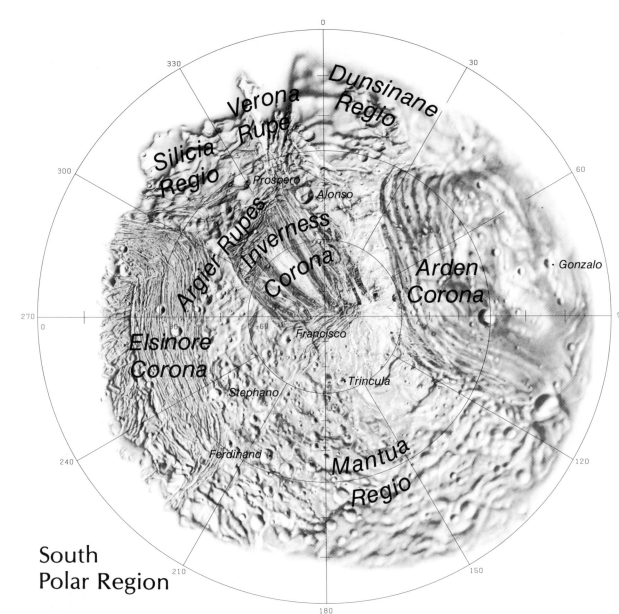

330

0

30

300

60

Dunsinane
Regio

Verona
Rupe

Silicia
Regio

270
0

90

Prospero

Alonso

Argier Rupes

Inverness
Corona

Arden
Corona

· Gonzalo

Francisco

Elsinore
Corona

-60

· Trincula

Stephano

240

120

Ferdinand

Mantua
Regio

210

150

180

South
Polar Region

Above: The Inverness Corona, once known as
'the Chevron,' is a series of pressure faults; the
Elsinore Corona, once known as 'the Race-
track,' is a series of parallel cracks; and the
Arden Corona, once known as 'the Pancake
Stack,' is a plateau whose edges have collapsed repeatedly. The nicknames were originally
applied informally when these features were
observed for the first time in January 1986. By
the end of the year they had been replaced pro-
visionally by more conventional nomenclature.

50   0   KILOMETERS   50   100

-0

-30

-60

-90

MIRANDA

Of the five large Uranian moons, Miranda is the smallest
and innermost, being less than three times larger than
1985 U1. It was discovered in 1948 by the Netherlands-born
American astronomer Gerard Kuiper (1906–1973), and is named
for the daughter of Prospero in Shakespeare's *The Tempest.*

Miranda's composition is about half water ice, with the balance being divided between silicate rock and methane-related
organic compounds. On its surface there are huge fault canyons
12 miles deep and evidence of intense geologic activity. It has
been much more geologically active than the other Uranian
moons.

Given Miranda's surface temperature, −335 degrees Fahrenheit, much of the moon's geologic activity must be the result of
the tidal effect of the gravitational pull of Uranus, which has
'mobilized' a flow of icy material at low temperature from within Miranda.

Data for Miranda
Diameter: 217 miles (150 km)
Distance from Uranus: 43,600 miles (80,000 km)
Rotational period (Mirandan day): 1.41 Earth days
Sidereal period (Mirandan year): 1.41 Earth days
Inclination to ecliptic plane: 0°
Mean surface temperature: −335° F

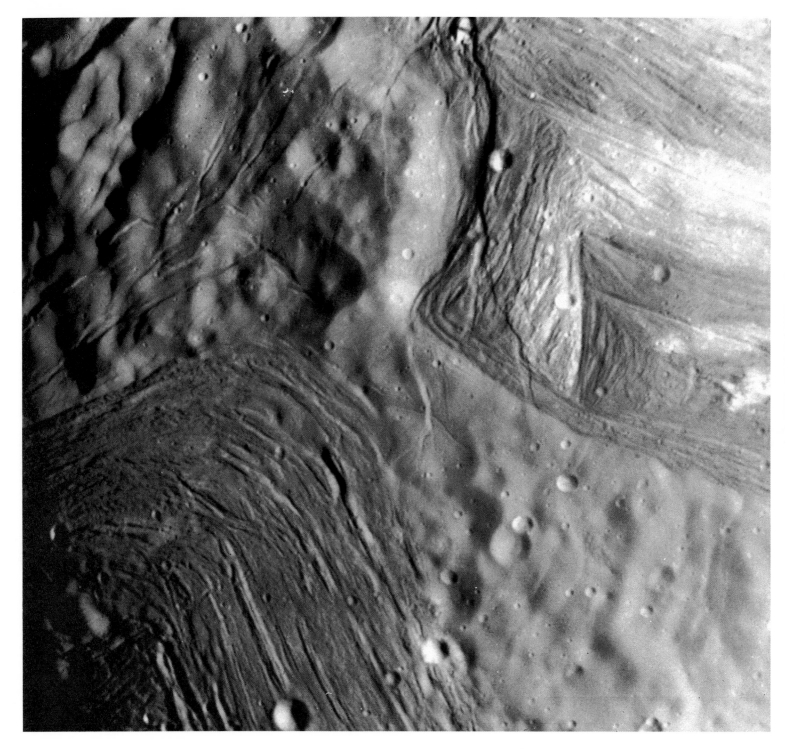

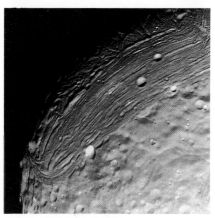

At left: This spectacular Voyager 2 photo shows the point on Miranda's surface at which Elsinore Corona (left) and Inverness Corona (right) nearly intersect.

At top: A mosaic image of Miranda, showing (left to right) Elsinore Corona, Inverness Corona and Arden Corona. The 'nick' at the top of this photo represents missing information.

Above: A view northwest over the Elsinore Corona.

Facing page photo: A Voyager 2 view over Inverness Corona toward Argier Rupes, with crater Alonso on the upper right.

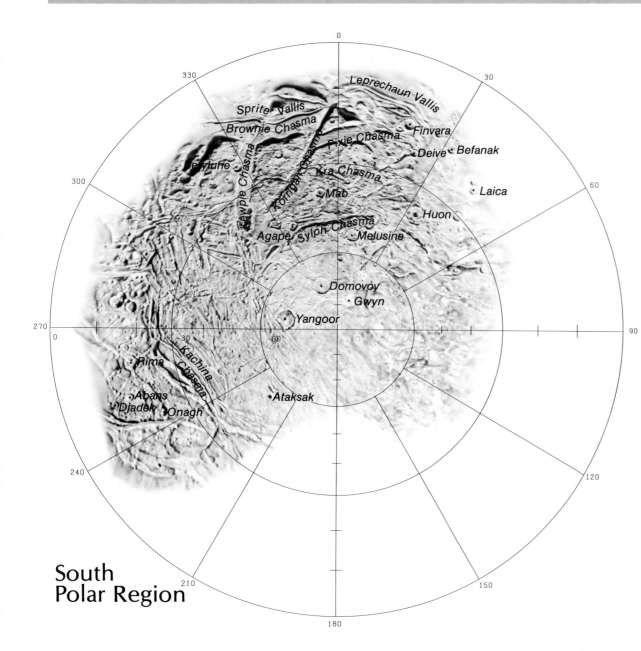

0
330
30
Leprechaun Vallis
Sprite Vallis
Brownie Chasma
Finvara
Pixie Chasma
Deive · Befanak
Kewpie Chasma
Kra Chasma
Korrigan Chasma
· Laica
300
· Mab
Sylph Chasma
· Huon
Agape
· Melusine
270
· Domovoy
· Gwyn
· Yangoor
90
-30
60
Kachina Chasma
Rima
· Abans
Djadek · Onagh
· Ataksak
240
120
210
South
Polar Region
150
180

The southern hemispheric map (*above*) of the bright moon Ariel coordinates nicely with the photo (*at right*) of the same area of Ariel. Recognizable on the lower left of the photographic moon is Kachina Chasmata, and at top center we find the complex of geographical features that includes Korrigan Chasma, Pixie Chasma and Brownie Chasma.

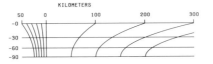

KILOMETERS
50 0   100   200   300
-0
-30
-60
-90

ARIEL

Discovered in 1851 by the English amateur astronomer William Lassell (1799–1880), Ariel is named for a spirit, the servant of Prospero, in *The Tempest* by William Shakespeare. Ariel's composition is about 50 percent water ice, 30 percent silicate rock and 20 percent methane ice. The major surface characteristic of Ariel are swaths of what appears to be fresh frost.

Ariel is largely devoid of impact craters with diameters in excess of 30 miles and has the brightest surface of any Uranian moon. Ariel also appears to have undergone a period of intense geologic activity, which has produced many fault canyons and has resulted in outflows of water ice from its interior. Where the longer canyons intersect, the surfaces are smooth, indicating that the valley floors are covered with huge glaciers.

Data for Ariel

Diameter: 720.8 miles (1160 km)
Distance from Uranus: 118,358 miles (190,900 km)
Rotational period (Arielian day): 2.52 Earth days
Sidereal period (Arielian year): 2.52 Earth days
Inclination to ecliptic plane: 0°

UMBRIEL

Discovered in 1851 by William Lassell (1799–1880) at the same time that he identified Ariel, Umbriel is named for the dusky sprite in Alexander Pope's *The Rape of the Lock*.

While the surfaces of all the Uranian moons are darkened by the presence of methane ice, Umbriel is the darkest. Even its impact craters, which should theoretically show lighter-colored water ice in their bottoms, are dark. Nevertheless, Umbriel is thought to be mostly composed of water ice with the balance made up of silicate rock and methane ice. Umbriel is like the other Uranian moons, but it carries most of its methane ice on its surface. Overall, it is the least geologically active of the Uranian moons.

Data for Umbriel

Diameter: 739.4 miles (1190 km)
Distance from Uranus: 165,284 miles (266,000 km)
Rotational period (Umbrielan day): 4.14 Earth days
Sidereal period (Umbrielan year): 4.14 Earth days
Inclination to ecliptic plane: 0°

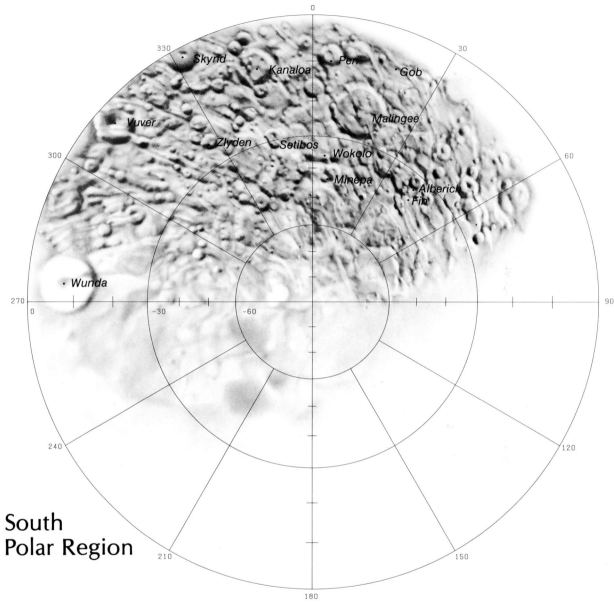

South
Polar Region

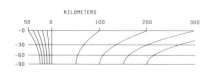

KILOMETERS

In this photograph (*at left*) relating to this map (*above*) of Umbriel, the craters (from the top) Skynd, Vuver and Wunda can be seen on the upper perimeter of the moon's left crescent. The light patch visible above these appears to be water ice, in contrast to the moon's overall dusky methane ice coating.

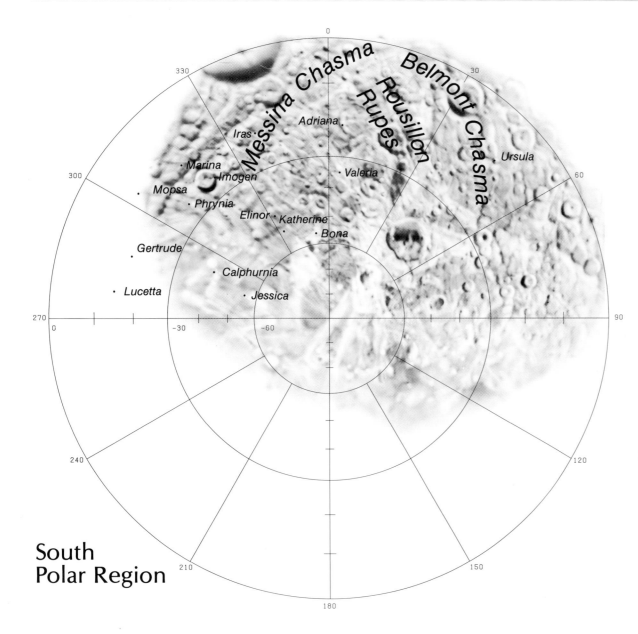

South
Polar Region

In co-ordinating this map (*above*) with the photo (*at right*) of Titania, Messina Chasmata, Belmont Chasmata and Rousillon Rupes can be seen as they arc down from the moon's upper crescent. Messina Chasmata is a huge rift canyon, and the large crater Gertrude can be seen at the left edge of the crescent.

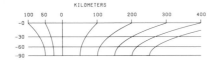

TITANIA

The largest of the Uranian moons, Titania was discovered in 1787 by William Herschel (1738–1822) six years after he discovered Uranus itself. Not to be confused with Titan, Saturn's largest moon, Titania is named for the fairy queen of medieval folklore who was the wife of Oberon.

Like its companions in the Uranian moon system, Titania is half composed of water ice, and like Ariel and Oberon, its surface is mostly water ice. The major feature on this icy surface is a huge canyon that dwarfs the scale of the Grand Canyon on Earth and is in a class with the Valles Marineris on Mars and the Ithaca Chasma on Saturn's Tethys.

Like the other Uranian moons, Titania is about 30 percent silicate rock and 20 percent methane-related organic compounds. It has been theorized that Titania may have a small co-orbital moon. If so, it would be the first co-orbital to be identified outside Saturn's system and it would be the 11th moon for Uranus.

Data for Titania

Diameter: 998.2 miles (1610 km)
Distance from Uranus: 271,104 miles (436,300 km)
Rotational period (Titanian day): 8.71 Earth days
Sidereal period (Titanian year): 8.71 Earth days
Inclination to ecliptic plane: 0°

OBERON

The outermost of the Uranian moons, Oberon was discovered by William Herschel (1738–1822) at the same time that he discovered Titania, and six years after his discovery of their mother planet. Oberon is named for the fairy king of Medieval folklore who was the husband of Titania. The name is derived from the Old French *Auberon* and is akin to the Old High German *Alberich,* which is thought to mean 'a white, ghostlike apparition.'

Like the other Uranian moons, Oberon is composed roughly of 50 percent water ice, 30 percent silicate rock and 20 percent methane-related carbon/nitrogen compounds. Unlike those of the others, however, Oberon's ancient, heavily-cratered surface shows very little evidence of internal geological activity. Being the second largest and most massive of the Uranian moons, as well as the farthest from the mother planet, Oberon is the moon least effected by the tidal effects of the gravity of Uranus.

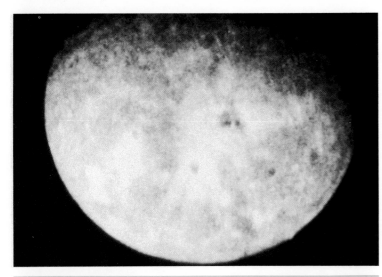

Data for Oberon

Diameter: 961 miles (1550 km)
Distance from Uranus: 362,507 miles (583,400 km)
Rotational period (Oberonian day): 13.46 Earth days
Sidereal period (Oberonian year): 13.46 Earth days
Inclination to ecliptic plane: 0°

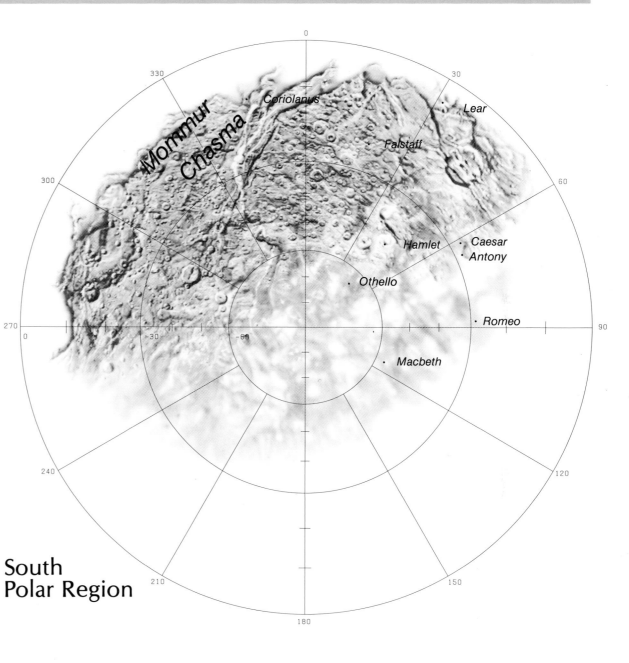

South
Polar Region

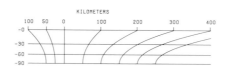

Above left: This photo of Oberon clearly correlates with the map *above*—half-hidden in shadow along the upper crescent, Mommur Chasma, crater Coriolanus (and its huge northern neighbor) and craters Falstaff and Lear are arranged left to right. Oberon is the outermost Uranian moon.

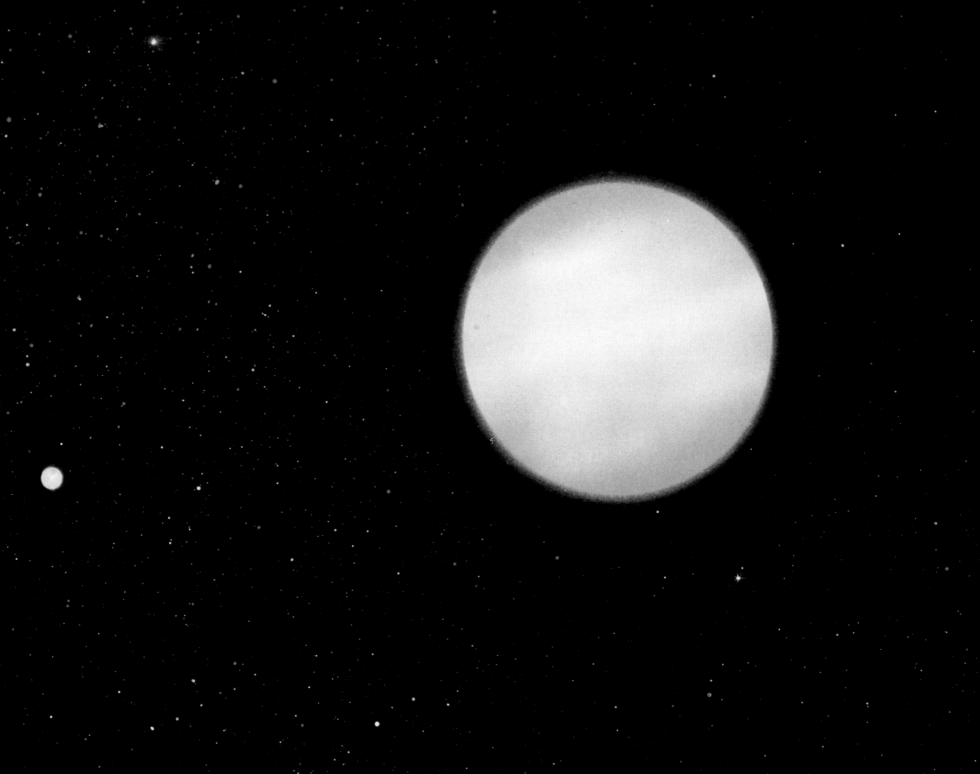

NEPTUNE

The eighth planet in the Solar System, Neptune was first observed by Galileo Galilei (1564–1642) in December 1612 and January 1613 while the great Italian astronomer was conducting his observations of Jupiter. Galileo mistakenly recorded it as a fixed star, despite his having observed its motion relative to other stars.

The existence of an eighth planet was predicted in 1845 by Urbain Jean Joseph LeVerrier (1811–1877) in France and John Crouch Adams (1819–1892) in England, based on their analysis of anomalies in the orbit of Uranus. The following year, Heinrich Ludwig D'Arrest (1822–1875) and Johann Galle physically observed it, and identified it as a planet for the first time.

Named for the Roman god of the sea, this blue-green giant is a near twin of Uranus in terms of its size and the chemical composition of its atmosphere. Unlike Uranus but *like* Jupiter, however, Neptune *radiates more heat than it absorbs* and it also has a greater density and mass than the slightly larger Uranus. Like Jupiter and Uranus, Neptune may eventually be shown to have a faint *ring system,* as has been suggested by occultation data.

Physically, Neptune is characterized by a dark band at its equator and lighter colored *temperate zones* in its northern and southern hemispheres. The dark band is perhaps evidence of the shadows of planetary rings or perhaps simply the absence in the equatorial region of the methane crystal haze and/or water ice crystal haze that drifts above the cloud tops in Neptune's colder temperate zones. Beneath this haze are clouds of ammonia and beneath these is a hydrogen *atmosphere.* Neptune probably has a rocky *core* roughly the size of the planet Earth that is covered to a depth of perhaps 5000 miles by an icy ocean of partly frozen water and liquid ammonia.

Nearly the twin of Uranus, Neptune departs from its 'brother' planet in that it partakes of the Jovian quality of radiating more energy than it receives. The illustration *on the facing page* is a NASA artist's visualization of Neptune and its largest moon, Triton.

Data for Neptune

Diameter: 30,758 miles (49,500 km)
Distance from Sun: 2,812,940,000 miles
(4,537,000,000 km) at aphelion
2,762,720,000 miles
(4,456,000,000 km) at perihelion
Mass: 4.672×10^{25} lb (1.028×10^{26} kg)
Rotational period (Neptunian day): 15.8 Earth hours
Sidereal period (Neptunian year): 60,189 Earth days
(165 Earth years)
Eccentricity: 0.009
Inclination of rotational axis: 29.56°
Inclination to ecliptic plane (Earth = 0): 1.77°
Albedo (100% reflection of light = 1): .34 to .5
Mean temperature: −343° F
Major atmospheric components: Hydrogen, Helium, Methane
Other atmospheric components: Ammonia, Argon

The time-sequential pictures of Neptune (shown *at right*) were taken at 9:40, 10:07, 10:19 and 11:14 universal time on 5 May 1979. The first, third and fourth pictures were recorded through an infrared methane absorption band filter, while the other picture shows the nonfiltered appearance of the planet in a spectral region which is relatively free of methane absorption.

In the later photos, high clouds of ice crystals in the northern and southern hemispheres produce two bright areas, while the absence of the haze layer in the equatorial region reveals a deeper layer of methane gas which shows here as the darker areas. During the period covered by these images the bright features may be seen to move towards the eastern limb of the planet.

The chart *below* shows Neptune's known moons—Triton and Nereid—in their orbital relation to the mother planet.

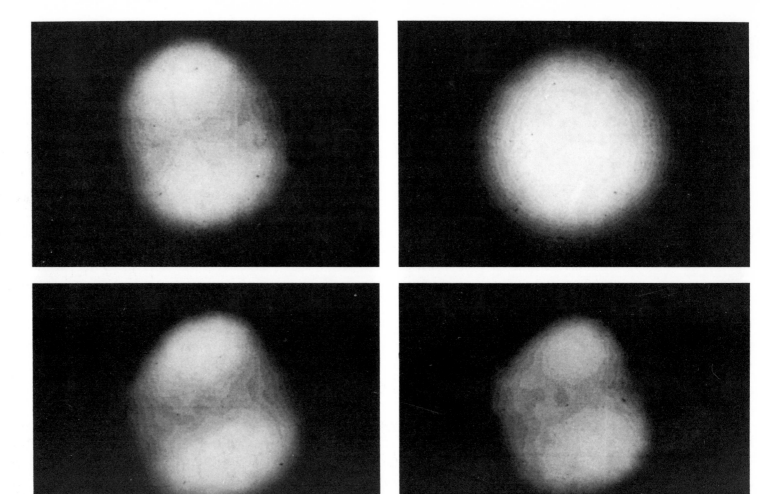

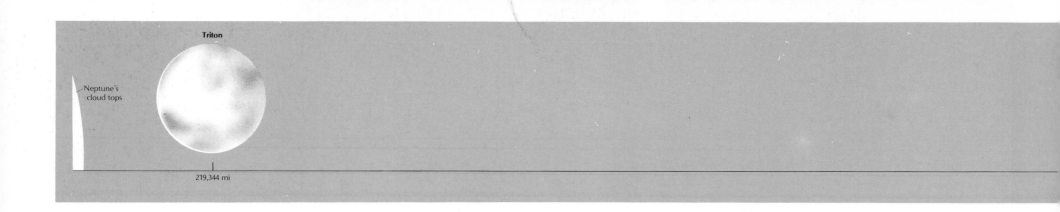

Triton

Neptune's cloud tops

219,344 mi

Prior to the American Voyager 2 spacecraft encounter with Neptune in 1989, the planet was known to possess two moons, although 1981 observations at the University of Arizona led to the prediction of at least one other moon. Prior to Voyager 2's encounter with Uranus in 1986, only five Uranian moons were known and Voyager's observations tripled that number. This fact alone would lead us to suspect that more Neptunian moons await discovery.

The two Neptunian moons known prior to 1986 are among the most peculiar in the Solar System. Triton, discovered less than a month after Neptune, is a huge object with the only *retrograde orbit* known in the Solar System while Nereid, discovered more than a century after Triton, has the most *elliptical orbit* of any known moon in the Solar System.

Data for Triton

Diameter: 3728 miles (6000 km)
Distance from Neptune: 219,344 miles (353,000 km)
Mass: 6.229×10^{22} lb (1.370×10^{23} kg)
Rotational period (Tritonian day): 5.87 Earth days
Sidereal period (Tritonian year): 5.87 Earth days
Inclination to ecliptic plane: 159.9°

Data for Nereid

Diameter: 310.7 miles (500 km)
Distance from Neptune: 3,454,823.8 miles (5,560,000 km)
Rotational period (Nereidan day): 359.9 Earth days
Sidereal period (Nereidan year): 359.9 Earth days
Inclination to ecliptic plane: 27.2°

TRITON

Neptune's largest moon was discovered by English brewer and amateur astronomer William Lassell (1799–1880) just 17 days after the discovery of Neptune itself. Triton, like its parent planet, is named for a god of the sea—in this case the merman son of the Greek god Poseidon and goddess Amphitrite.

Unique among all the known moons in the Solar System, Triton revolves around its mother planet in the direction opposite to Neptune's rotation. Its orbit is so close to Neptune, and is gradually getting so much closer, that one day Neptune's gravity might pull it apart and scatter it into a Saturn-like ring. (However, if this does happen it will not be for several million years.) Triton's surface, unlike that of Neptune, is rocky rather than gaseous—and this rocky surface is probably covered by methane frost and, perhaps, a faint methane *atmosphere.*

NEREID

Neptune's second moon was discovered in 1949 by the Dutch-American astronomer Gerard Peter Kuiper (1906–1973) in an elliptical orbit far beyond the orbit of Triton. Named for the Nereids—sea nymph daughters of the Greek god Nereis—Nereid is much tinier than its brother Triton. Little is known about Nereid other than its extremely *elliptical* orbit and its size relative to Triton.

Nereid

3,454,824 mi

PLUTO

During the mid to late nineteenth century, astronomers studying the revolutions of Uranus and Neptune detected slight anomalies that could only be explained by the gravitational effect of another body farther out in the Solar System. Around the turn of the century, Percival Lowell (1855–1916) took up a systematic search of the heavens, looking for what he called 'Planet X.' When Lowell died in 1916, others continued the search, including William Pickering (1858–1938) of Harvard, who called the yet-undiscovered object 'Planet O.' In 1915 and again in 1919, Pluto was actually *photographed, but not noticed* because it was much fainter than it had been predicted to be. By this time, the organized search for Planet X was largely abandoned. In the meantime, Pickering altered his theory regarding the hypothetical location of 'Planet O' and for the first time predicted that the perihelion of its orbit might actually bring it briefly closer to the Sun than Neptune. It was a radical idea that turned out to be accurate for Pluto.

In 1929 the Lowell Observatory at Flagstaff, Arizona resumed the search begun by its founder, using a 13-inch telescope and a wide-field survey camera. This proved to be the right approach, and on 18 February 1930, the young astronomer Clyde Tombaugh (1906–   ) identified a new planet in some photographs he had taken the previous month. The discovery was announced a month later on the 149th anniversary of the discovery of Uranus, and the new planet was called Pluto after the Roman god of the dead and the ruler of the underworld. The name was considered appropriate because of the planet's enormous distance from the Sun's warmth, and also because the first two letters were Percival Lowell's initials.

In the first years after it was discovered, physical data about Pluto was virtually impossible to obtain. In 1950 however, Gerard Kuiper at the Mount Palomar Observatory estimated its diameter at 3658 miles, making it the second smallest planet in the Solar System. In 1965, it was observed in occultation with a 15th magnitude star, confirming that its diameter could not exceed 4200 miles. Thus it was that the 3658 mile estimate held until the 1970s.

In 1976 methane ice was discovered to exist on Pluto's surface. Until then, the planet's faintness had been attributed to its being composed of dark rock. Since ice would tend to reflect light more so than dark rock, it would follow that if it *were* 3658 miles in diameter *and* covered with methane ice, it would be brighter than it is. Therefore, it was decided that Pluto was smaller than originally suspected, leading us to conclude that its diameter is less than the 2160 mile diameter of the Earth's Moon and probably as small as 1375 miles. This would make it the smallest of the nine planets and smaller than *seven* of the planetary moons. As estimates of Pluto's size continue to be revised downward, it becomes less and less likely that it has the mass to exert the gravitational force on Neptune's orbit that was originally predicted.

James Christy, who identified Charon in 1978 as an actual moon—and not a shape aberration of Pluto—pronounces the first syllable of the moon's name 'Char' as in Charlene—in honor of his wife, for whom the moon was named. This is just as Percival Lowell's name initially appears in the first two letters of the planet's name: Pl-uto!

In May of 1987, scientists at NASA's Jet Propulsion Laboratory (JPL) reported that observations of Pluto revealed an extensive atmosphere of natural gas surrounding the planet—which effectively fragments theories that Pluto is actually an asteroid.

As 82 year old Clyde Tombaugh—who discovered Pluto in his youth—put it, 'Many years ago I found that Pluto in fact looks like a planet.' Given that it is confirmed to have a moon, Tombaugh added that Pluto 'feels like a planet. Now, with JPL's confirmation of an atmosphere, it smells like a planet. It must be a planet.' And so it is. . . on the edge of the Solar System.

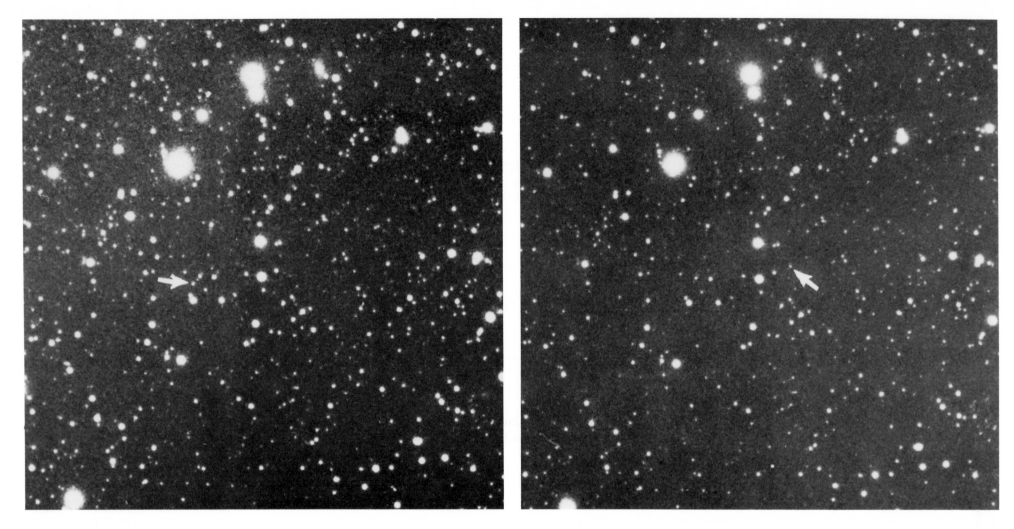

Above left and right: These are copies of the Pluto discovery plates which Clyde Tombaugh made of the Constellation Gemini on 23 and 29 January 1930 at Lowell Observatory in Flagstaff, Arizona. Comparing them on 18 February, Tombaugh made a startling discovery—one of the pricks of light had moved during those six days. A quick comparison to a 21 January plate confirmed that Tombaugh had discovered the long-predicted trans Neptunian planet!

As we know, the newly-discovered planet was named Pluto, and is indicated above by a white arrow—which indicates its sequential position in transit of the three longitudinally arranged stellar objects in the center of each photo.

If true, this would mean that Pluto is *not* 'Planet X,' and that 'Planet X' still exists and is yet to be discovered.

It has also been suggested that Pluto is perhaps the largest of a theorized belt of *trans-Neptunian asteroids.* However, that notion fails to take into account that Pluto is two and a half times the diameter of Ceres, the largest known asteroid, and nearly seven times larger than the average of the 18 largest known asteroids. Among the arguments that *can* be made for its not being a planet, or at least for its not being a 'normal' planet, are the aspects of its peculiar behavior. As we have noted, it has an extremely *elliptical orbit.* This orbit ranges from an aphelion of 49AU to a perihelion of 27AU. The latter is actually closer to the Sun than the perihelion of Neptune's much more circular

orbit, as Pickering had predicted. It has been pointed out that this highly elliptical orbit is characteristic of asteroids such as 944 Hidalgo and 2060 Chiron.

A second aspect of Pluto's behavior that sets it apart from other planets is its steep inclination to the ecliptic plane. The orbits of all the planets are within two and one-half degrees of this same plane except Mercury which is inclined at seven degrees and Pluto itself—which is inclined at an acute 17 degrees, making it very unusual among its peers.

A theory concerning the physical nature of Pluto holds that at one time it was actually one of the moons of Neptune. It is further theorized that Pluto was somehow thrown out of its Neptunian orbit by some calamitous interaction with Neptune's moon Triton—perhaps even a collision. One of the Solar System's largest moons, Triton is more than twice the size of Pluto, and as such might have had the gravitational force to slam a competing object out of Neptunian orbit if it ventured close enough. Both Triton and Neptune's other known moon Nereid have unusual orbits that might possibly be relics of such a colossal event.

While its behavior partially defines it, and certainly sets it apart from other planets, less is known about Pluto's physical characteristics than is known about any other planet. Since no spacecraft will visit it in the twentieth century, we are left only with educated guesses about Pluto. We know that it is extremely cold, with noontime summer temperatures rarely creeping above −350 degrees Fahrenheit. Its rocky surface is known to also contain methane, probably in the form of ice or frost. Water ice may also be present, though this is not likely, and Pluto's mass suggests a rocky *core*. Pluto has generally been thought to have no *atmosphere*, because its relatively small mass wouldn't give it sufficient gravity to retain an atmosphere, and it is too cold for even such substances as methane to easily exist in their gaseous state. However, Scott Sawyer of the U of Texas has discovered what may be a tenuous methane vapor atmosphere on Pluto.

Data for Pluto

Diameter: 1375 miles (2200 km)
Distance from Sun: 4,572,500,000 miles
(7,375,000,000 km) at aphelion
2,743,500,000 miles
(4,424,000,000 km) at perihelion
Mass: 3×10^{23} lb (6.6×10^{23} kg)
Rotational period (Plutonian day): 6.3 Earth days (6 days, 9 hrs, 18 min) (retrograde)
Sidereal period (Plutonian year): 90,465 Earth days
(248 Earth years)
Eccentricity: 0.250
Inclination of rotational axis: 50°
Inclination to ecliptic plane (Earth = 0): 17.2°
Albedo (100% reflection of light = 1): .5
Mean surface temperature: −382.27° F
Maximum surface temperature: −350° F
Minimum surface temperature: −390° F
Major atmospheric components: Methane

CHARON

The discovery of the Plutonian moon Charon came about indirectly in 1978. While James Christy at the US Naval Observatory in Flagstaff, Arizona was attempting to measure Pluto's size, he thought he'd noticed that it was not spherical. Further observations led him to the conclusion that the elongation he had observed was due to the presence of a satellite very close to Pluto. Further calculations indicated that this newly discovered body was as close as 10,563 miles from Pluto.

Pluto's moon, named Charon—after the mythological son of Erebus and Nox who was appointed by the gods to ferry the souls of the dead across the river Styx—is closer in size to its mother planet than any other moon in the Solar System. Estimates of its diameter range from 497 to 744 miles, the later being roughly half the diameter of Pluto. Its mass has been estimated at between five and 10 percent that of Pluto. Little is known about its physical properties, although it is possible that methane ice exists on Charon as it does on Pluto.

Charon revolves around Pluto every 153 hours, exactly matching Pluto's rotational period, meaning that the same hemisphere of Charon faces the same hemisphere of Pluto at all times. (From Earth, we can observe only one side of the Moon's surface, but the Earth rotates against the revolution of the Moon so that the Moon is visible regularly from most regions of the Earth's surface.)

Data for Charon

Diameter: Possibly 1200 km
Distance from Pluto: 12,028 miles (19,400 km)
Rotational period (Charonian day): 6.3 Earth days
(6 days, 9 hrs, 18 min)
Sidereal period (Charonian year): 6.3 Earth days
Inclination to ecliptic plane: Unknown
Mean surface temperature: −382° F

COMETS

The most etherial of the objects in the Solar System, comets are essentially 'snowballs' of carbon dioxide, methane or water ice that have extremely *elliptical orbits* and which exhibit spectacular *tails* when heated by the Sun. Comets travel around the Sun in fixed elliptical orbits that have perihelions less than that of the inner terrestrial planets and aphelions that can be as great or greater than the outermost planets. The period of time that it takes for a comet to make a complete revolution around the Sun may range from a few years to many centuries. The comets Encke and Giacobini-Zinner, for example, are observed with periods of 3.3 and 6.5 years respectively, while Halley's Comet has an observed period of 76.3 years. Some comets have periods of truly extraordinary durations. The comet Kohoutek which was discovered in 1973 will not make another turn around the Sun until the year 76,973 and the Great Comet of 1864 will not reappear for 2.8 million years!

Because of their small size, comets cannot be seen from Earth until the Sun heats up their enormous tails. Thus, they have in years past seemed to appear suddenly and almost from nowhere. In Roman times they were considered 'bad omens' and in the Middle Ages their appearance was always considered to be some sort of 'sign.' The arrival overhead of Halley's Comet in 1066, however, was certainly not a 'bad omen' for William the Conqueror at the Battle of Hastings.

In 1682, Edmond Halley (1656–1742), England's second Astronomer Royal successfully and scientifically calculated the period of the Great Comet of 1682 at 76.3 years. He did not live to see it reappear in 1758, but it did reappear as he predicted it would, and it has borne his name—and has been the most fam-ous of comets—ever since. Since Halley's time we have understood comets as scientifically predictable natural phenomena rather than as 'omens' or 'signs.' Nevertheless, comet Halley's 1910 appearance led to numerous predictions of catastrophe, and even of the end of the world. It is uncertain what sort of mumbo-jumbo might be read into the fact that Halley's Comet was very difficult to see from Earth with the naked eye in 1986.

Above: A Medieval portrait of a comet as seen above a European hamlet. In those days comets were not understood and were considered to be omens.

Facing page: This spectacular object is none other than Halley's Comet, photographed during its celebrated 1910 visit to the near-Earth region of the Solar System. The most well-known of comets, Halley's was the center of enormous media attention in 1910.

Famous Comets

Date	Comet	Period (years)
1066	Halley's Comet	76.3
1811	Great Comet of 1811	3000
1815	Olbers' Comet	74.0
1819	Encke's Comet	3.3
1819	Pons-Winnecke Comet	6.0
1843	Great Comet of 1843	512.4
1844	Great Comet of 1844	102,050
1858	Donati's Comet	2040 (?)
1864	Great Comet of 1864	2,800,000
1889	Swift's Second Comet	7.0
1892	Holmes' Comet	6.9
1900	Comet Giacobini-Zinner	6.5
1923	d'Arrest's Comet	6.6
1925	Comet Schwassmann-Wachmann	16.2
1957	Comet Mrkos	5.3
1973	Comet Kohoutek	75,000

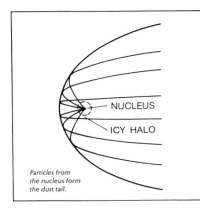

Particles from the nucleus form the dust tail.

Above: This is a basic diagram of the components of a comet.

Above right: The upper diagram here was drawn from a perspective parallel to the ecliptic plane; and the lower diagram was drawn from a perspective perpendicular to the ecliptic plane.

These are diagrammatic models of the comet Giacobini-Zinner at perihelion, or its point of nearest approach to the Sun.

Spanning these pages is a photograph—taken on 15 January 1974 at the University of Arizona's Catalina Observatory—of the much-heralded Comet Kohoutek, whose appearance in our celestial neighborhood was fraught with dire predictions, and thus was reminiscent of ancient cometary visits in human history.

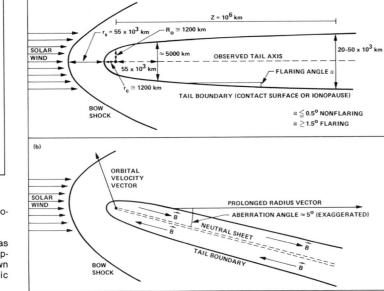

A comet, as we've noted, is basically an icy 'snowball.' This 'snowball,' usually containing a solid core, is the permanent part of the comet and is known as its *nucleus*. The cometary nucleus is usually quite small. Halley's Comet, for example, has an *irregular nucleus* that is only about nine miles long and five miles wide. The nucleus is in turn surrounded by a hydrogen cloud.

As the nucleus nears the orbit of Mars on its journey inward from the distant reaches of the Solar System, interaction with the solar warmth causes it to develop a fuzzy halo or *coma*. (The word *coma* as used here is the Latin word for 'hair,' rather than the Greek word for 'deep sleep.') As the nucleus and coma reach the orbit of the Earth, a tail begins to develop as the solar wind blows material away from the nucleus. Upon the formation of the tail, the coma is known as the comet's *head*. Tails may sometimes be composed of particulate impurities released from the nucleus, in which case they will appear curved. Tails may also be composed of ionized gases, in which case they will appear to be straight.

Cometary tails grow longer and brighter the closer they approach the Sun. Some may be almost invisible, and some, such as that of Halley's in 1910, may span half the visible sky! Donati's Comet, which appeared in 1858, had *two* gas tails and a *particulate dust* tail that reached a length of nearly 50 million miles. The Great Comet of 1811 had a coma with a diameter of 1.2 million miles and had a tail nearly 100 million miles in length. The longest tail on record, however, was that of the Great Comet of 1843 which reached from near the Sun to well beyond Mars, a distance of over 200 million miles!

Some comets may get no closer to the Sun than the orbit of Jupiter or Mars, but some, such as 1979-XI, get so close that they collide with the Sun and are destroyed. There is no confirmed instance of a comet colliding with the Earth, but in 1908 an object glowing brighter than the Sun, and suspected to have been a comet, crashed into the Tunguska region of Siberia, doing massive damage.

After a comet rounds the Sun and starts its journey back to the outer Solar System, the tail, still solar wind-blown ions or particles, is observed to *precede* the nucleus rather than to *trail* it. As the comet moves away from the Sun, the tail grows smaller and finally vanishes. The coma then shrinks and disappears and the barely visible icy nucleus recedes into the far reaches of the Solar System.

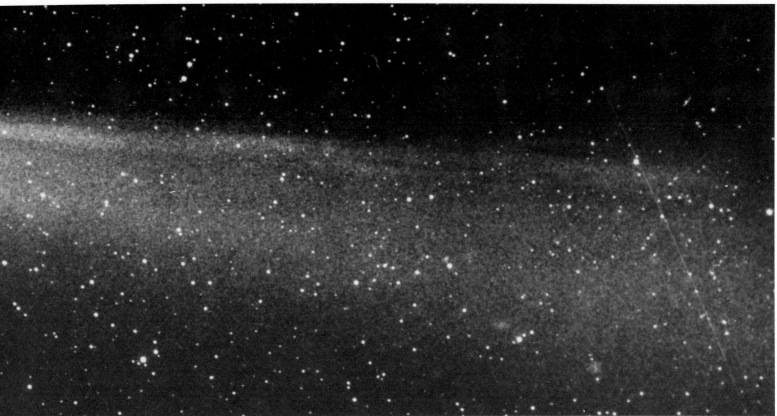

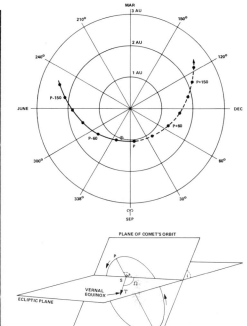

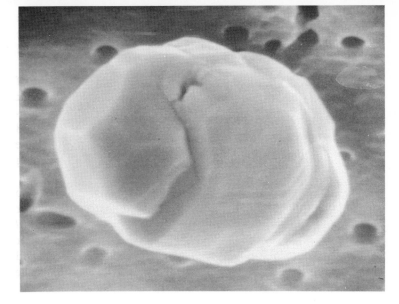

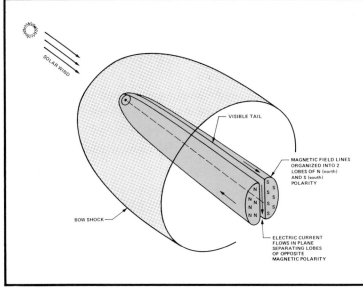

Below far left: An electron micrograph of a suspected cometary dust particle.

At left: An artist's conception of the basic mechanisms at work in a comet's interaction with the solar wind, which produces a visible comet tail composed of magnetic lobes. This model is based upon the comet Giacobini-Zinner.

A comet's path toward and away from perihelion is shown *at top, above:* The crosshairs sign represents the Earth, and the minus numbers represent the comet's approach toward perihelion, just as the plus numbers represent its movement back out into the farther reaches of the Solar System. The comet's position is noted in 30-day intervals.

Above: A diagram defining the orbital angle of the Giacobini-Zinner comet, with 'i' being the actual measured angle (relative to the ecliptic plane).

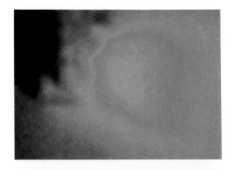

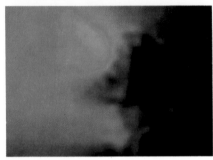

Above: A series of images of the icy nucleus of Halley's Comet, photographed in 1986 by the European Space Agency spacecraft Giotto, taken from distances of 5900 and 3100 miles.

Below: Graphs representing the aphelion and perihelion of Halley's extremely eliptical orbit.

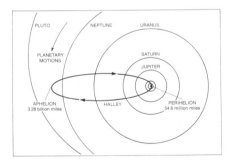

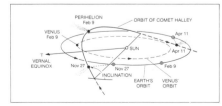

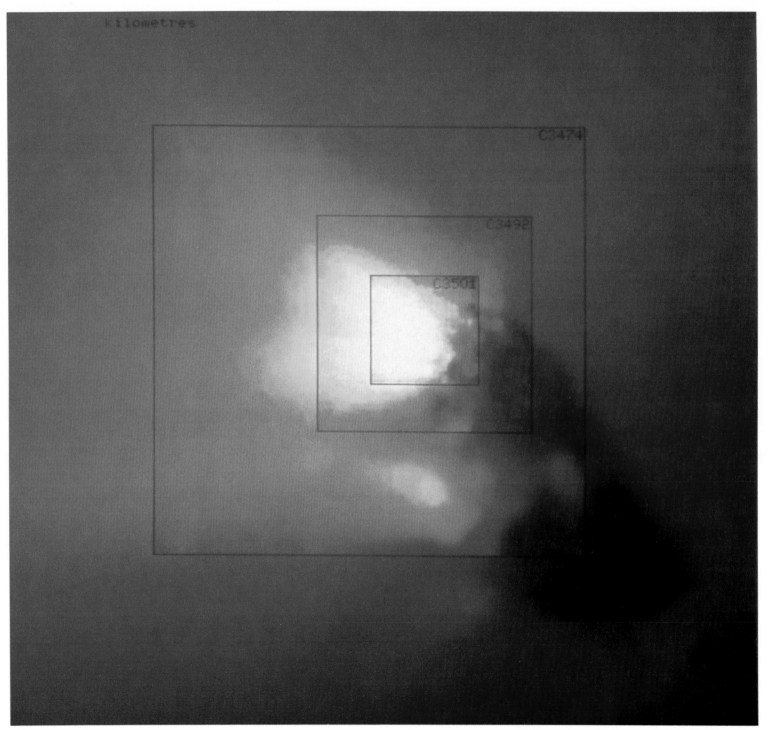

Above: Comet Iketa-Seki appears to streak toward the horizon, showing its classic tail.

At left: This Giotto photo very clearly shows Halley's dented, oblong shaped nucleus, with dust jets escaping from the sunlit side of the comet.

Facing page: This Giotto-generated composite image of the comet's nucleus was taken from such a position that north is at the top of the photo and the Sun is located off-camera and to the left. Note the dust jets escaping from the nucleus in the innermost frame. Also note the gas envelope surrounding the nucleus.

METEORITES

Known as *meteoroids* when outside a planet's atmosphere, as *meteors* when falling through the Earth's atmosphere and as *meteorites* when they're found intact or when referring to their craters, these relatively small bodies exist throughout the Solar System and frequently impact other bodies in the Solar System. Despite their small size, meteorites are an extremely important component of the Solar System because their impact craters have contributed extensively to the surface texture of nearly every known planet, moon and asteroid; the major meteorite cratering took place in the few million years that followed the formation of the Solar System 4.6 billion years ago, but the effects remain. Most planets and moons have thousands of four billion year-old meteorite impact craters, many of incredible size. On the Earth's Moon there are Copernicus and Tycho, and the incredible eroded craters such as those that formed Mare Crisium and Mare Orientale. Jupiter's Callisto was battered by so many meteorites that virtually no other type of surface can be seen, while on Saturn's Mimas the crater Herschel spans a third of that moon's diameter.

Not only have meteorites created the features by which we recognize various bodies in the Solar System, they provide us with the means to determine the sequence of events in a body's history. For example, craters that have been disturbed by faulting, erosion or subsequent impact craters can be identified as relatively ancient, while those that are superimposed on other features can be dated as more recent. The relative number of craters found on or near another type of geologic feature can serve to date that feature. For example, Mare Imbrium on the Earth's Moon is a huge, ancient crater that was formed very

soon after the Moon itself. This basin was in turn filled by an outflowing of molten rock 3.2 billion years ago and the number of *new* craters in this smoother basin, compared to the number of craters on older terrain, helps to date the epoch when the Solar System received most of its meteorite cratering.

The most intense period of cratering took place in the Solar System's first half billion years and, though it has tapered off, it has never stopped. On Earth most, but not all, of the effects of cratering have been eradicated by erosion. Only on Jupiter's Io, which is constantly being resurfaced by frenetic volcanic activity, is there a visible surface unscarred by meteorite craters.

While most of the extremely large meteoroids were expended in the Solar System's first half billion years, the Solar System's moons and planets have been subjected to a constant bombardment of smaller particles for the subsequent four billion years. On planets and moons without atmospheres, meteoroids simply strike the surface full force. In the case of planets and moons with atmospheres, however, all but the very largest burn up in the atmosphere, briefly becoming fiery-tailed meteors or *shooting stars*—the very bright ones being called *fireballs*.

In the Earth's atmosphere, meteoroids become visible as meteors at an altitude of 60 miles. The fiery streaks are due to

Above: This photomosaic of Mercury illustrates a terrain which is extensively scarred by meteorite impacts—as would be the case on any planet lacking an atmosphere. Mercury does not have sufficient gravity to hold an atmosphere (which would serve to incinerate objects such as meteors before they could impact the surface). Note here that many of the craters themselves have been scarred by other meteorites, thus yielding the effect of craters upon craters.

Dates	Meteor Showers	Constellation
1-4 Jan	Quadrantids	Boötes
5-10 Feb	Alpha Aurigids	Auriga
10-12 March	Zeta Boötids	Boötes
19-23 April	Lyrids	Hercules
1-6 May	May Aquarids	Aquarius
30 May	Eta Pegasids	Pegasus
27-30 June	Pons-Winnecke meteors	Draco
14 July	Alpha Cygnids	Cygnus
26-31 July	Delta Aquarids	Aquarius
10-14 Aug	Perseids	Cassiopeia
10-20 Aug	Kappa Cygnids	Cygnus
22 Sept	Alpha Aurigids	Auriga
2 Oct	Quadrantids	Boötes
9 Oct	Giacobinids	Draco
18-23 Oct	Orionids	Orion
14-18 Nov	Leonids	Leo
10-13 Dec	Geminids	Gemini

Above: This meteorite was discovered in 1981 by an Antarctic meteorite recovery team and is thought to be of possible Martian origin.

At right: This meteorite is kept in a dry nitrogen cabinet for study in the meteorite processing lab in Johnson Space Center's Planetary and Earth Sciences Laboratory. Designated as sample 79001, it was found in Antarctica and is thought to be of Martian origin.

compression, rather than friction—as is occasionally suggested. Since meteoroids enter the Earth's atmosphere at speeds of up to 45,000 mph, the air ahead of them is compressed like the air in the cylinder of a diesel engine and the heat is transferred to the moving body, turning it into a meteor or shooting star.

There are thousands of meteors entering the Earth's atmosphere every year, but their average weight is less than one ounce and even the brightest fireballs weigh less than five pounds and are burned up without reaching the Earth's surface. More than 2000 meteorites have been discovered on the surface, however, and eight have been identified as weighing more than 15 tons. The largest, the Hoba West discovered near Grootfontein in Southwest Africa, weighs 70 tons. The second largest, Ahnighito, was found by Admiral Peary at Cape York, Greenland and weighs 34 tons.

One of the largest, and certainly one of the most well preserved, meteorite craters on Earth is located in northeastern Arizona. Known as Barringer Crater, the crater is similar to those on Mercury or the Moon. An impressive sight when viewed firsthand, it is 3900 feet across and 600 feet deep, but is hardly a large crater by Solar System standards.

Meteorites on Earth are classified geologically as *Aerolites* (stone), *Siderites* (stony iron), *Siderolites* (iron) and *Tektites* (volcanic glass)—although the extraterrestrial origin of Tektites

Above: This meteorite weighing 21.6 pounds was recovered near Lost City, Oklahoma on 9 January 1970, following its entry into the Earth's atmosphere on 3 January. Prior to entry it is estimated that the meteor weighed approximately one ton.

At left: This is one of Earth's most famous meteorite scars, the Barringer Crater in Arizona. It is 3900 feet wide and 600 feet deep. Compare this to the craters depicted in the Earth's Moon, Mercury and Mars sections of this book.

is now in doubt. All of the largest meteorites, such as the Hoba West and Ahnighito, are Siderolites, while the largest Aerolite on record (found in China in 1976) weighed only 3894 pounds. Siderolites are the most common meteorites and Tektites are the rarest. While other types of meteorites are found throughout the Earth and have been falling to the surface continuously for 4.6 billion years, only four major Tektite showers are theorized to have occurred in history. These were in Europe and North America during the *Tertiary Period* and in Australia and Africa during the *Pleistocene Period.*

Meteorites provide enigmatic views of the geology of the Solar System because they are frequently composed of rock types unlike anything that exists on Earth. It is thought that some meteorites may have had their origin in other planets or within the Asteroid Belt.

Meteorites fall to Earth during *meteor showers* which occur as the Earth passes through the highly elliptical orbital path of a particular cluster of meteorites. Meteors can be observed nearly every clear night, and *major* meteor showers occur 18 times per year, or 18 times in the Earth's revolution around the Sun. These showers, as noted in the accompanying table, are identified as originating in a particular constellation. They do not of course originate *in* the constellation, just as meteors are not really shooting *stars* but rocks that have their origin in our own Solar System. Extremely spectacular *Leonid meteor showers* occur in November every 33 years. In 1833 they were recorded as raining down 'like snowflakes.' They were also recorded as especially profuse in 1799, 1866 and 1966. The 1899 and 1933 showers were no more vivid than the typical annual Leonid shower because of an apparent aberration of their orbit, caused when they crossed the orbits of Jupiter and Saturn. It is suggested, however, that one mark one's calendar for the November 1999 event.

ASTEROIDS

In the 342 million mile interval between the orbits of Mars and Jupiter is a vast collection of small planet-like objects called *asteroids*. While their name translates as implying a star-like character, the asteroids are more accurately described as *planetoids* or minor planets. Literally they are fragments of rock that may have their origin in the cataclysmic destruction of one or several terrestrial planets, or they may be debris left over from the origin of the Solar System itself. The largest asteroid, Ceres, .is 485 miles in diameter, but there are only six known asteroids with diameters greater than 100 miles.

The discovery of the field of minor planets that we know as the *Asteroid Belt*, dates to the theoretical work of German astronomer Johann Elert Bode (1747–1826) of the Berlin Observatory.

In 1772 Bode authored Bode's Law, which took into account the regular intervals between the known planets and postulated that a planet should, by his law, exist between Mars and Jupiter. Little did Bode realize that this interval was not filled by a single planet, but by thousands of planetoids. In 1800 Bode's countryman, Johann Hieronymus Schroeter (1745–1816) organized what he called the Celestial Police, an association of astronomers dedicated to finding the planet whose existence had been postulated by Bode (and by Titius of Wittenberg before him).

Ironically the first asteroid, Ceres, was not discovered by a member of Schroeter's Celestial Police, but was discovered on New Years Day in 1801 by Giuseppe Piazzi (1746–1826), director of the observatory at Palermo, Sicily. Though Piazzi later joined

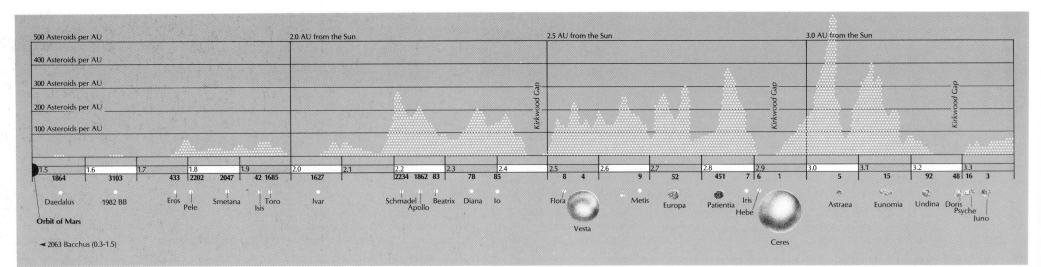

the Celestial Police, the most successful member of the 'force' would have to be Heinrich Olbers (1748–1840) a German amateur astronomer who was able to 'recover' or rediscover Ceres in 1802. Olbers then went on to discover the asteroids Pallas and Vesta in 1802 and 1807, respectively. It would be over 30 years, however, before a fifth asteroid would be discovered.

In 1830, another German amateur astronomer, Karl Ludwig Hencke (1793–1866) went on a search for further asteroids which finally bore fruit in 1845 with the discovery of Astraea. Two years later, Hencke discovered a sixth asteroid, Hebe. By the 1840s photography was brought into play as a tool, and suddenly the search for new asteroids took on a whole new flavor. Both Iris and Flora were discovered in 1847, the same year that Hencke found Hebe, and after that several were found each year. Within 10 years, 48 asteroids had been discovered and by 1899 there were 451 known asteroids. By 1930, the year that Clyde Tombaugh discovered Pluto, there were more than 1000 known asteroids. After World War II, the International Astronomical Union set up a co-operative program of asteroid research. By 1980 there were more than 2000 known asteroids, and by 1986 the number exceeded 3450.

Because it is now possible to detect smaller and smaller asteroids, the number of known asteroids is likely to increase indefinitely. The total number of asteroids is estimated at 30,000, but theoretically they can range down to sand grain-sized specks that will probably never all be counted.

With the flurry of discovery in the nineteenth century, the convention was adopted to number the asteroids in the order they were discovered (it is also conventional to incorporate the assigned numeral into the official name). Because of differences in magnitude owing to composition and distance from Earth, the numbers denoting order of discovery don't necessarily list the

The Fifteen Largest Asteroids

Size Order	Number	Name	Diameter (mi)	Diameter (km)	Year Discovery
1	1	Ceres	621.86	1003	1801
2	2	Pallas	376.96	608	1802
3	4	Vesta	333.56	538	1807
4	10	Hygeia	279	450	1849
5	31	Euphrosyne	229.4	370	1854
6	704	Interamnia	217	350	1910
7	511	Davida	200:26	323	1903
8	65	Cybele	191.58	309	1861
9	52	Europa	179.18	289	1858
10	451	Patienta	171.12	276	1899
11	15	Eunomia	168.64	272	1851
12	16	Psyche	155	250	1851
	48	Doris	155	250	1857
	92	Undina	155	250	1867
13	324	Bamberga	152.52	246	1892
14	24	Themis	145.08	234	1853
15	95	Arethusa	142.6	230	1867

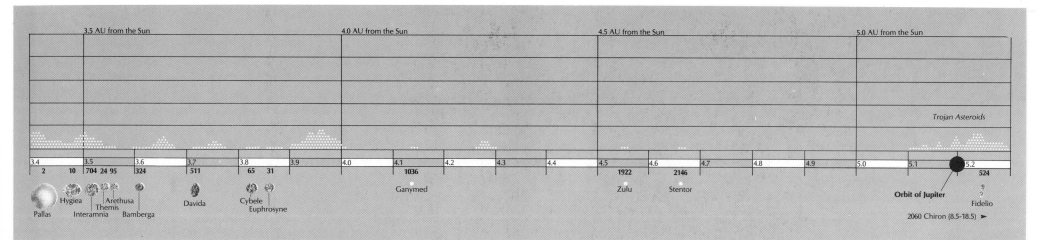

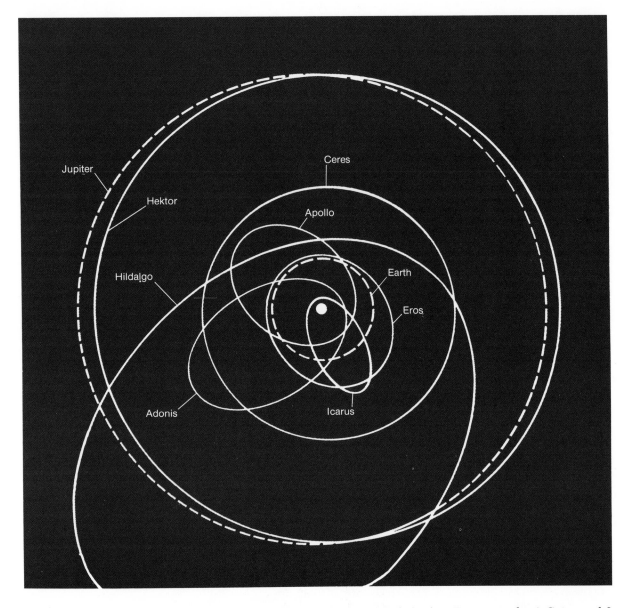

Above: This is a chart of some of the asteroid orbits which lie between the Sun and Jupiter. Note the highly elliptical orbits of asteroids Hidalgo, Apollo, Adonis and Icarus. Icarus, as can be seen, is especially aptly named for the mythological youth whose fall resulted from flying too close to the Sun.

the Roman goddess of the hearth, 5 Astraea for the Greek goddess of justice, 6 Hebe for the Greek goddess of youth and spring, 7 Iris for the Greek messenger and goddess of the rainbow, while 8 Flora was named for the Roman goddess of flowers. Since the mid-nineteenth century, however, the naming convention has been stretched to include, among the classical goddesses, such names as those of astronomers and Earthly geographical locations. Today asteroids can be named after anything, usually at the prerogative of the discoverer. In some cases asteroids have been assigned names already assigned to other objects in the Solar System. For example, the asteroids 52 Europa and 1036 Ganymede co-exist with the Jovian moons Europa and Ganymede. Some asteroids have simply been assigned unromantic code numbers, but recently named asteroids also include 2309 Mr Spock (formerly 1971 QX²), named for a character on the *Star Trek* television series, and the asteroids numbered 3350 through 3356, which were renamed in March 1986 for the seven persons killed two months earlier in the accident involving the American spacecraft *Challenger*.

Few asteroids have anything approximating the staid near-circular orbits of most planets. In fact, asteroid orbits are extremely elliptical. It is for this reason that we have chosen to use *Astronomical Units (AU)* rather than miles/kilometers for the tables in this section. Astronomical Units (1 AU = 93 million miles) can be used as a sort of shorthand to allow us to easily grasp the immensity of the eccentricity of these orbits without an overwhelming blizzard of digits. While 1 Ceres has an eccentric orbit that varies between 2.55 AU and 2.94 AU from the Sun, 944 Hidalgo varies between 2.00 AU and 9.61 AU, and the amazing 2060 Chiron has a perihelion of 18.50 AU and an aphelion of 8.50 AU. This translates as a difference of 930 million miles between its closest and furthest approaches to the Sun!

While the greatest concentration of asteroids is in the Asteroid Belt between Mars and Jupiter, there are a number of interesting exceptions to this rule, and 2060 Chiron is only one. In 1898, it was discovered that the eccentric orbit of 433 Eros brought it within the orbit of Mars, and in 1931 and 1975 it came within .15 AU of the Earth (1.15 AU from the Sun). By the mid-twentieth century, a fairly large number of asteroids were found to cross the orbit of Mars. These include 1566 Icarus and 1862 Apollo, which actually cross the *Earth's* orbit and come as close to the Sun as .19 AU and .65 AU respectively! The names for these two are well chosen. Apollo was the Greek god of the Sun, while Icarus was the character from Greek mythology who attempted to escape from Crete using homemade wings, but who crashed

asteroids in the order of their size. For example, 1 Ceres and 2 Pallas are the two largest asteroids, but the third largest is 4 Vesta, while 3 Juno is actually sixth in size order.

The naming of the first asteroids followed the same pattern of classical mythology used with the planets and their moons: 1 Ceres was named for the tutelary goddess of Sicily, 2 Pallas for the goddess Pallas Athena, 3 Juno for the sister of Jupiter and the prinicipal goddess of the Roman pantheon, 4 Vesta for into the sea when the wax on his wings melted because he came *too close to the sun.* After the discovery of 1862 Apollo in 1932, it was 'lost' for 41 years. After its rediscovery in 1973, the name Apollo has been applied as a class name to *all* asteroids whose elliptical orbits cross the orbit of the Earth.

While 2060 Chiron, whose aphelion of 8.50 AU never even brings it within the orbit of Jupiter, is hardly in the Apollo class, it opens up a whole area of speculation about asteroids because of its incredible eccentricity and the fact that it never comes near the Asteroid Belt. Some astronomers have suggested that it might actually be a tailless comet nucleus, while others have suggested that it might be the harbinger of a whole new Asteroid Belt awaiting discovery in the distant reaches of the Solar System.

Another interesting class of asteroids are the Trojan asteroids which exist at the outer edge of the Asteroid Belt and, in fact, travel in the same orbital path as Jupiter. Travelling at the speed of the Jovian revolution around the Sun, a group of *Leading Trojans* precede Jupiter in its orbit by 60 degrees, while *Trailing Trojans* follow at the same speed, remaining 60 degrees behind.

Some of the asteroids are observed to be characterized by other interesting peculiarities. At least two of them, 2 Pallas and 12 Victoria, actually have smaller satellite asteroids orbiting around them like moons around a planet! The 'moon' associated with 2 Pallas bears the same size relationship to 2 Pallas that the Moon does to the Earth. Meanwhile, 8 Flora has a mass of at least 13 tiny bodies that accompany it in its orbit, and five of these have diameters in excess of 18 miles.

Asteroids vary widely in their shapes and other characteristics.

Most of the well-known larger ones such as 1 Ceres, 3 Juno and 4 Vesta are roughly spherical or just slightly oblique, like 2 Pallas. Many, however, are as eccentric in their shapes as they are in their orbits: some, such as 41 Daphne, 44 Mysa, 107 Camilla and 349 Dembowska are so elongated as to appear stretched and almost cigar-like.

One very oddly shaped asteroid is 624 Hektor. One of the Trojan asteroids sharing Jupiter's orbit, 624 Hektor might be a dogbone-shaped object, or it may actually be *two* asteroids. If it is the latter, it would be the only known *binary* asteroid and as such, the two may actually be touching one another. If 624 Hektor *is* a contact binary, 624A and 624B would probably be identical in size and mass.

Asteroids are also varied in their color and composition, with the very dark carbonaceous *C-types* accounting for three-quarters of the known asteroids. The C-types include 1 Ceres, 2 Pallas, 10 Hygiea, 65 Cybele and 624 Hektor. Roughly 15 percent of known asteroids are the rust red *S-types,* rich in iron and magnesium silicates. These include 3 Juno, 6 Hebe, 7 Iris and 15 Eunomia, as well as 8 Flora and its associated family. About five percent of the asteroids are classed as *M-types,* with characteristics suggestive of metallic substances. The remaining asteroids vary widely from 4 Vesta, which is composed of light grey basalt, to the metallic 16 Psyche, to the bright and olivine-rich 349 Dembowska.

Because of their size, none of the asteroids have the mass to permit the formation of an atmosphere or the more complex features noted on planets, but they are potentially rich in data about the origin of the Solar System.

THE FIRST TEN ASTERIODS

Number	Name	Absolute Magnitude	Discovery		(AU)	Farthest Distance from Sun (Miles)	(Km)	(AU)	Closest Distance to Sun (Miles)	(Km)	Diameter (Miles)	(Km)	Sidereal Period in Earth years	Albedo
1	Ceres	4.5	Piazzi	1 Jan 1801	2.94	273,420,000	440,206,200	2.55	237,150,000	381,811,500	621.86	1003	4.60	.054
2	Pallas	5.0	Olbers	23 Mar 1802	3.42	318,060,000	512,076,600	2.11	196,230,000	315,920,300	376.96	608	4.61	.074
3	Juno	6.5	Harding	1 Sep 1804	3.35	311,550,000	501,595,500	1.98	184,140,000	296,465,400	155.00	250	4.36	.151
4	Vesta	4.3	Olbers	29 Mar 1807	2.57	239,010,000	384,806,100	2.15	199,950,000	321,919,500	333.56	538	3.63	.229
5	Astraea	8.1	Hencke	8 Dec 1845	3.06	284,580,000	458,173,800	2.10	195,300,000	314,433,000	62.54	117	4.14	.140
6	Hebe	7.0	Hencke	1 Jul 1847	2.92	271,560,000	437,211,600	1.93	179,490,000	288,978,900	120.90	195	3.78	.164
7	Iris	6.8	Hind	13 Aug 1847	2.94	273,420,000	440,206,200	1.84	171,120,000	275,503,200	129.58	209	3.68	.154
8	Flora	7.7	Hind	18 Oct 1847	2.55	237,150,000	381,811,500	1.86	172,980,000	278,497,800	93.62	151	3.27	.144
9	Metus	7.8	Graham	25 Apr 1848	2.68	249,240,000	401,276,400	2.09	194,370,000	312,935,700	93.62	151	3.69	.139
10	Hygiea	6.5	DeGasparis	12 Apr 1849	3.46	321,780,000	518,065,800	2.84	264,120,000	425,233,200	279.00	450	5.60	.041

THE OTHER ASTEROIDS

No	Name	Absolute Magnitude	Discovery
11	Parthenope	7.8	DeGasparis 11 May 1850
12	Victoria	8.4	Hind 13 Sep 1850
13	Egeria	8.1	DeGasparis 12 Nov 1850
14	Irene	7.5	Hind 19 May 1851
15	Eunomia	6.4	DeGasparis 29 Jul 1851
16	Psyche	6.9	DeGasparis 17 Mar 1852
17	Thetis	9.1	Luther 17 Apr 1852
18	Melpomene	7.7	Hind 24 June 1852
19	Fortuna	8.4	Hind 22 Aug 1852
20	Massilia	7.7	DeGasparis 19 Sep 1852
21	Lutetia	8.6	Goldschmidt 15 Nov 1852
22	Calliope	7.3	Hind 16 Nov 1852
23	Thalia	8.2	Hind 15 Dec 1852
24	Themis	8.3	DeGasparis 5 Apr 1853
25	Phocea	9.3	Charcornac 6 Apr 1853
26	Proserpina	8.8	Luther 5 May 1853
27	Euterpe	8.4	Hind 8 Nov 1853
28	Bellona	8.2	Luther 1 Mar 1854
29	Amphitrite	7.1	Marth/Pogson 1 Mar 1854
30	Urania	8.8	Hind 22 Jul 1854
31	Euphrosyne	7.3	Ferguson 1 Sep 1854
32	Pomona	8.8	Goldschmidt 26 Oct 1854
33	Polyhymnia	9.9	Charcornac 28 Oct 1854
34	Circe	9.6	Charcornac 6 Apr 1855
35	Leukothea	9.7	Luther 19 Apr 1855
36	Atalanta	9.8	Goldschmidt 5 Oct 1855
37	Fides	8.4	Luther 5 Oct 1855
38	Leda	9.7	Charcornac 12 Jan 1856
39	Laetitia	7.4	Charcornac 8 Feb 1856
40	Harmonia	8.3	Goldschmidt 1 Mar 1856
41	Daphne	8.2	Goldschmidt 22 May 1856
42	Isis	8.8	Pogson 23 May 1856
43	Ariadne	9.2	Pogson 15 Apr 1857
44	Nysa	7.8	Goldschmidt 27 May 1857
45	Eugenia	8.3	Goldschmidt 28 Jun 1857
46	Hestia	12.5	Pogson 16 Aug 1857
47	Aglaia	12.9	Luther 15 Sep 1857
48	Doris	12.1	Goldschmidt 19 Sep 1857
49	Pales	12.7	Goldschmidt 19 Sep 1857
50	Virginia	13.6	Ferguson 4 Oct 1857
51	Nemausa	11.2	Laurent 22 Jan 1858
52	Europa	11.7	Goldschmidt 4 Feb 1858
53	Calypso	13.1	Luther 4 Apr 1858
54	Alexandra	12.2	Goldschmidt 11 Apr 1858
55	Pandora	12.1	Scarle 10 Sep 1858
56	Melete	9.5	Goldschmidt 9 Sep 1857
57	Mnemosyne	8.4	Luther 22 Sep 1859
58	Concordia	9.9	Luther 24 Mar 1860
59	Elpis	8.8	Charcornac 13 Sep 1860
60	Echo	10.0	Ferguson 15 Sep 1860
61	Danaë	8.9	Goldschmidt 9 Sep 1860
62	Erato	9.8	Lesser 14 Sep 1860
63	Ausonia	9.0	DeGasparis 11 Feb 1861
64	Angelina	8.8	Tempel 6 Mar 1861
65	Cybele	8.0	Tempel 10 Mar 1861
66	Maia	10.5	Tuttle 10 Apr 1861
67	Asia	9.7	Pogson 18 Apr 1861
68	Leto	8.2	Luther 19 Apr 1861
69	Hesperia	8.2	Sciaparelli 29 Apr 1861
70	Panopea	8.9	Goldschmidt 5 May 1861
71	Niobe	8.3	Luther 13 Aug 1861
72	Feronia	10.1	Peters 29 Jan 1862
73	Clytie	10.3	Tuttle 7 Apr 1862
74	Galatea	10.0	Tempel 29 Aug 1862
75	Eurydice	10.0	Peters 22 Sep 1862
76	Freia	9.1	D'Arrest 14 Nov 1862
77	Frigga	9.7	Peters 12 Nov 1862
78	Diana	9.2	Luther 15 Mar 1863
79	Eurynome	9.2	Watson 14 Sep 1863
80	Sappho	9.2	Pogson 3 May 1864

Data for the Asteroids listed below includes Number, Name and Absolute Magnitude:

No	Name	Absolute Magnitude
81	Terpsichore	9.6
82	Alkmene	9.5
83	Beatrix	9.8
84	Klio	10.3
85	Io	8.9
86	Semele	9.7
87	Sylvia	8.1
88	Thisbe	8.1
89	Julia	8.1
90	Antiope	9.4
91	Aegina	10.0
92	Undina	7.9
93	Minerva	8.7
94	Aurora	8.7
95	Arethusa	8.8
96	Aegle	9.3
97	Klotho	8.7
98	Ianthe	9.9
99	Dike	11.2
100	Hekate	9.1
101	Helena	9.5
102	Miriam	10.3
103	Hera	8.8
104	Klymene	9.4
105	Artemis	9.4
106	Dione	8.8
107	Camilla	8.3
108	Hecuba	9.7
109	Felicitas	10.1
110	Lydia	8.7
111	Ate	9.1
112	Iphigenia	10.9
113	Amalthea	9.9
114	Kassandra	9.5
115	Thyra	8.8
116	Sirona	8.9
117	Lomia	9.2
118	Peitho	10.1
119	Althaea	9.8
120	Lachesis	8.8
121	Hermione	8.5
122	Gerda	9.2
123	Brunhild	10.1
124	Alkeste	9.4
125	Liberatrix	9.8
126	Velleda	10.6
127	Johanna	9.5
128	Nemesis	8.8
129	Antigone	7.8
130	Elektra	8.5
131	Vala	11.0
132	Aethra	10.3
133	Cyrene	9.2
134	Sophrosyne	9.6
135	Hertha	9.2
136	Austria	10.8
137	Meliboea	9.1
138	Tolosa	11.0
139	Juewa	9.2
140	Siwa	9.6
141	Lumen	9.6
142	Polana	11.6
143	Adria	10.5
144	Vibilia	9.1
145	Adeona	8.7
146	Lucina	9.3
147	Protogeneia	9.9
148	Gallia	8.5
149	Medusa	11.9
150	Nuwa	9.3
151	Abundantia	10.5
152	Atala	9.6
153	Hilda	8.8
154	Bertha	8.5
155	Scylla	12.5
156	Xanthippe	9.8
157	Dejanira	12.4
158	Koronis	10.8
159	Aemilia	9.3
160	Una	10.1
161	Athor	10.4
162	Laurentia	10.0
163	Erigone	10.8
164	Eva	9.8
165	Loreley	8.4
166	Rhodope	10.9
167	Urda	10.6
168	Sibylla	9.3
169	Zelia	10.7
170	Maria	10.7
171	Ophelia	9.4
172	Baucis	10.1
173	Ino	8.8
174	Phaedra	9.6
175	Andromache	9.6
176	Iduna	9.5
177	Irma	10.8
178	Belisana	10.7
179	Klytaem-nestra	9.3
180	Garumna	11.6
181	Eucharis	9.1
182	Elsa	10.2
183	Istria	11.0
184	Dejopeja	9.5
185	Eunike	8.7
186	Celuta	10.5
187	Lamberta	9.5
188	Menippe	10.5
189	Phthia	10.8
190	Ismene	8.7
191	Koiga	10.0
192	Nausikaa	8.6
193	Ambrosia	10.9
194	Prokne	8.8
195	Eurykleia	10.1
196	Philomela	7.7
197	Arete	10.8
198	Ampella	9.5
199	Byblis	10.0
200	Dynamene	9.5
201	Penelope	9.3
202	Chryseis	9.1
203	Pompeja	10.1
204	Kallisto	10.1
205	Martha	10.0
206	Hersilia	9.8
207	Hedda	11.1
208	Lacrimosa	10.5
209	Dido	9.5
210	Isabella	10.5
211	Isolda	9.0
212	Medea	9.3
213	Lilaea	10.1
214	Aschera	10.4
215	Oenone	10.9
216	Kleopatra	8.2
217	Eudora	11.0
218	Bianca	10.0
219	Thusnelda	10.7
220	Stephania	12.3
221	Eos	8.9
222	Lucia	10.4
223	Rosa	11.2
224	Oceana	9.8
225	Henrietta	9.8
226	Weringia	11.0
227	Philosophia	10.1
228	Agathe	14.0
229	Adelinda	10.5
230	Athamantis	8.6
231	Vindobona	10.6
232	Russia	11.5
233	Asterope	9.6
234	Barbara	10.4
235	Carolina	9.8
236	Honoria	9.5
237	Coelestina	10.8
238	Hypatia	9.2
239	Adrastea	11.8
240	Vanadis	9.9
241	Germania	8.6
242	Kriemhild	10.7
243	Ida	11.2
244	Sita	13.4
245	Vera	9.3
246	Asporina	10.0
247	Eukrate	9.3
248	Lameia	11.3
249	Ilse	12.3
250	Bettina	8.5
251	Sophia	11.3
252	Clementina	10.7
253	Mathilde	11.4
254	Augusta	13.2
255	Oppavia	11.4
256	Walpurga	11.1
257	Silesia	10.2
258	Tyche	9.5
259	Aletheia	9.0
260	Huberta	10.3
261	Prymno	10.6
262	Valda	12.9
263	Dresda	11.8
264	Libussa	9.7
265	Anna	12.5
266	Aline	9.5
267	Tirza	11.8
268	Adorea	9.8
269	Justitia	11.0
270	Anahita	10.0
271	Penthesilea	11.0
272	Antonia	11.9
273	Atropos	12.0
274	Philagoria	11.2
275	Sapientia	10.0
276	Adelheid	9.7
277	Elvira	11.2
278	Paulina	10.6
279	Thule	9.8
280	Philia	11.9
281	Lucretia	13.1
282	Clorinde	12.0
283	Emma	9.8
284	Amalia	11.3
285	Regina	11.9
286	Iclea	10.2
287	Nephthys	9.6
288	Glauke	11.1
289	Nenetta	10.6
290	Bruna	13.2
291	Alice	12.8
292	Ludovica	11.0
293	Brasilia	11.1
294	Felicia	11.3
295	Theresia	11.4
296	Phaëtusa	14.0
297	Caecilia	10.4
298	Baptistina	12.3
299	Thora	13.0
300	Geraldina	10.8
301	Bavaria	11.4
302	Clarissa	12.0

303	Josephina	10.1
304	Olga	10.8
305	Gordonia	10.3
306	Unitas	10.0
307	Nike	10.8
308	Polyxo	9.3
309	Fraternitas	11.6
310	Margarita	11.7
311	Claudia	11.2
312	Pierretta	10.2
313	Chaldaea	10.1
314	Rosalia	10.9
315	Constantia	14.0
316	Goberta	11.5
317	Roxane	11.3
318	Magdalena	10.3
319	Leona	11.6
320	Katharina	11.8
321	Florentina	11.4
322	Phaeo	10.3
323	Brucia	11.1
324	Bamberga	8.1
325	Heidelberga	9.8
326	Tamara	10.3
327	Columbia	11.4
328	Gudrun	10.3
329	Svea	10.8
330	Adalberta	14.0
331	Etheridgea	10.7
332	Siri	10.8
333	Badenia	10.4
334	Chicago	8.6
335	Roberta	9.9
336	Lacadiera	11.0
337	Devosa	9.9
338	Budrosa	9.8
339	Dorothea	10.5
340	Eduarda	11.5
341	California	12.6
342	Endymion	11.3
343	Ostara	12.8
344	Desiderata	9.1
345	Tercidina	10.1
346	Hermentaria	8.9
347	Pariana	10.1
348	May	10.7
349	Dembowska	7.2
350	Ornamenta	9.4
351	Yrsa	10.3
352	Gisela	11.6
353	Ruperto-Carola	12.3
354	Eleonora	7.5
355	Gabriella	11.6
356	Liguria	9.3
357	Ninina	9.8
358	Apollonia	10.3
359	Georgia	10.5
360	Carlova	9.4
361	Bononia	9.6
362	Havnia	10.1
363	Padua	10.0
364	Isara	11.1
365	Corduba	10.3
366	Vincentina	9.8
367	Amicitia	12.1
368	Haidea	11.1
369	Aëria	9.6
370	Modestia	11.7
371	Bohemia	10.0
372	Palma	8.5
373	Melusina	10.3
374	Burgundia	10.1
375	Ursula	8.6
376	Geometria	10.7
377	Campania	10.0
378	Holmia	11.2
379	Huenna	10.0
380	Fiducia	10.6
381	Myrrha	9.7
382	Dodona	9.8
383	Janina	11.0
384	Burdigala	10.8
385	Ilmatar	8.8
386	Siegena	8.6
387	Aquitania	8.4
388	Charybdis	9.5
389	Industria	9.4
390	Alma	11.4
391	Ingeborg	12.3
392	Wilhelmina	10.9
393	Lampetia	9.3
394	Arduina	10.9
395	Delia	11.5
396	Aeolia	11.1
397	Vienna	10.5
398	Admete	12.0
399	Persephone	10.3
400	Ducrosa	11.4
401	Ottilia	10.3
402	Chloë	10.3
403	Cyane	10.6
404	Arsinoë	10.0
405	Thia	9.8
406	Erna	11.5
407	Arachne	10.3
408	Fama	10.8
409	Aspasia	8.3
410	Chloris	9.5
411	Xanthe	10.2
412	Elisabetha	10.3
413	Edburga	11.3
414	Liriope	10.8
415	Palatia	10.5
416	Vaticana	9.2
417	Suevia	10.6
418	Alemannia	11.0
419	Aurelia	9.5
420	Bertholda	9.4
421	Zähringia	13.0
422	Berolina	11.8
423	Diotima	8.6
424	Gratia	10.8
425	Cornelia	10.8
426	Hippo	9.8
427	Galene	10.7
428	Monachia	13.0
429	Lotis	10.8
430	Hybris	11.9
431	Nephele	10.0
432	Pythia	10.3
433	Eros	12.4
434	Hungaria	12.4
435	Ella	11.3
436	Patricia	11.2
437	Rhodia	11.6
438	Zeuxo	10.9
439	Ohio	10.8
440	Theodora	13.0
441	Bathilde	9.4
442	Eichsfeldia	11.0
443	Photographica	11.5
444	Gyptis	9.1
445	Edna	10.3
446	Aeternitas	10.2
447	Valentine	10.4
448	Natalie	11.4
449	Hamburga	11.0
450	Brigitta	11.5
451	Patientia	7.7
452	Hamiltonia	13.4
453	Tea	12.0
454	Mathesis	10.3
455	Bruchsalia	9.9
456	Abnoba	11.1
457	Alleghenia	13.0
458	Hercynia	10.4
459	Signe	11.7
460	Scania	11.9
461	Saskia	11.4
462	Eriphyla	10.8
463	Lola	12.8
464	Megaira	10.5
465	Alekto	11.0
466	Tisiphone	9.4
467	Laura	12.1
468	Lina	10.6
469	Argentina	10.0
470	Kilia	11.4
471	Papagena	7.8
472	Roma	10.4
L473	Nolli	
474	Prudentia	11.7
475	Ocllo	12.4
476	Hedwig	9.8
477	Italia	11.5
478	Tergeste	9.2
479	Caprera	10.9
480	Hansa	10.0
481	Emita	9.9
482	Petrina	10.4
483	Seppina	9.7
484	Pittsburghia	11.5
485	Genua	9.8
486	Cremona	12.2
487	Venetia	9.4
488	Kreusa	8.9
489	Comacina	9.5
490	Veritas	9.4
491	Carina	10.0
492	Gismonda	11.1
493	Griseldis	11.7
494	Virtus	10.1
495	Eulalia	11.7
496	Gryphia	13.1
497	Iva	10.7
498	Tokio	10.3
499	Venusia	10.5
500	Selinur	10.6
501	Urhixidur	10.1
502	Sigune	11.7
503	Evelyn	10.4
504	Cora	11.3
505	Cava	10.1
506	Marion	9.7
507	Laodica	10.6
508	Princetonia	9.4
509	Iolanda	9.7
510	Mabella	11.0
511	Davida	7.4
512	Taurinensis	11.9
513	Centesima	10.8
514	Armida	10.6
515	Athalia	11.8
516	Amherstia	9.4
517	Edith	10.4
518	Halawe	12.5
519	Sylvania	10.4
520	Franziska	12.2
521	Brixia	9.8
522	Helga	10.1
523	Ada	10.8
524	Fidelio	11.0
525	Adelaide	14.0
526	Jena	11.0
527	Euryanthe	11.5
528	Rezia	10.2
529	Preziosa	11.2
530	Turandot	10.3
531	Zerlina	12.3
532	Herculina	7.0
533	Sara	11.0
534	Nassovia	10.9
535	Montague	10.6
536	Merapi	9.3
537	Pauly	9.9
538	Friederike	10.6
539	Pamina	11.0
540	Rosamunde	12.2
541	Deborah	11.4
542	Susanna	10.6
543	Charlotte	10.7
544	Jetta	11.3
545	Messalina	9.7
546	Herodias	11.0
547	Praxedis	10.9
548	Kressida	12.8
549	Jessonda	11.7
550	Senta	10.5
551	Ortrud	10.5
552	Sigelinde	10.7
553	Kundry	13.6
554	Peraga	9.8
555	Norma	11.7
556	Phyllis	10.4
557	Violetta	13.2
558	Carmen	10.1
559	Nanon	10.7
560	Delila	11.8
561	Ingwelde	12.3
562	Salome	11.0
563	Suleika	10.0
564	Dudu	11.5
565	Marbachia	12.3
566	Stereoskopia	9.2
567	Eleutheria	10.4
568	Cheruskia	10.6
569	Misa	11.3
570	Kythera	10.2
571	Dulcinea	12.9
572	Rebekka	11.8
573	Recha	10.8
574	Reginhild	13.8
575	Renate	12.4
576	Emanuela	11.0
577	Rhea	10.8
578	Happelia	10.7
579	Sidonia	9.1
580	Selene	10.9
581	Tauntonia	10.8
582	Olympia	10.2
583	Klotilde	10.3
584	Semiramis	9.8
585	Bilkis	11.5
586	Thekla	10.3
587	Hypsipyle	13.6
588	Achilles	9.7
589	Croatia	10.0
590	Tomyris	11.3
591	Irmgard	11.8
592	Bathseba	10.8
593	Titania	10.3
594	Mireille	14.0
595	Polyxena	9.3
596	Scheila	10.0
597	Bandusia	10.5
598	Octavia	10.8
599	Luisa	9.5
600	Musa	11.4
601	Nerthus	10.5
602	Marianna	9.4
603	Timandra	13.6
604	Tekmessa	10.5
605	Juvisia	10.6
606	Brangäne	11.6
607	Jenny	10.9
608	Adolfine	11.8
609	Fulvia	11.3
610	Valeska	13.3
611	Valeria	10.8
612	Veronika	12.4
613	Ginevra	11.1
614	Pia	12.1
615	Roswitha	11.4
616	Elly	11.8
617	Patroclus	9.1
618	Elfriede	9.4
619	Triberga	11.3
620	Drakonia	12.4
621	Werdandi	11.8
622	Esther	11.6
623	Chimaera	12.2
624	Hektor	8.6
625	Xenia	11.6
626	Notburga	10.3
627	Charis	11.2
628	Christine	10.3
629	Bernardina	10.8
630	Euphemia	12.4
631	Philippina	10.2
632	Pyrrha	13.4
633	Zelima	11.2
634	Ute	11.0
635	Vundtia	10.3
636	Erika	10.8
637	Chrysothemis	12.0
638	Moira	11.0
639	Latona	9.3
640	Brambilla	10.1
641	Agnes	13.7
642	Clara	11.3
643	Scheherezade	10.9
644	Cosima	11.7
645	Agrippina	11.0
646	Kastalia	14.3
647	Adelgunde	12.7
648	Pippa	10.9
649	Josefa	14.2
650	Amalasuntha	13.6
651	Antikleia	11.2
652	Jubilatrix	12.6
653	Berenike	10.5
654	Zelinda	9.5
655	Briseïs	10.7
656	Beagle	11.0
657	Gunlöd	12.0
658	Asteria	11.7
659	Nestor	9.8
660	Crescentia	10.6
661	Cloelia	10.8
662	Newtonia	11.8
663	Gerlinde	10.5
664	Judith	11.1
665	Sabine	9.7
666	Desdemona	12.0
667	Denise	10.4
668	Dora	13.4
669	Kypria	11.4
670	Ottegebe	10.9
671	Carnegia	11.5
672	Astarte	12.6
673	Edda	11.5
674	Rachele	8.6
675	Ludmilla	9.3
676	Melitta	10.6
677	Aaltje	11.0
678	Fredegundis	10.1
679	Pax	10.4
680	Genoveva	10.7
681	Gorgo	11.9
682	Hagar	13.4
683	Lanzia	9.5
684	Hildburg	11.9
685	Hermia	12.9
686	Gersuind	10.9
687	Tinette	12.8
688	Melanie	11.7
689	Zita	13.3
690	Wratislavia	8.8
691	Lehigh	10.6
692	Hippodamia	10.3
693	Zerbinetta	10.2
694	Ekard	10.1
695	Bella	9.9
696	Leonora	10.3
697	Galilea	10.8
698	Ernestina	12.0
699	Hela	13.2
700	Auravictrix	12.2
701	Oriola	10.4
702	Alauda	8.3
703	Noëmi	13.7
704	Interamnia	7.2
705	Erminia	9.5
706	Hirundo	12.1
707	Steïna	14.0
708	Raphaela	11.9
709	Fringilla	10.0
710	Gertrud	12.3
711	Marmulla	13.2
712	Boliviana	9.3
713	Luscinia	9.9
714	Ulula	10.3
715	Transvaalia	11.2
716	Berkeley	12.0
717	Wisibada	12.0
718	Erida	10.9
L719	Albert	
720	Bohlinia	10.7
721	Tabora	10.5
722	Frieda	13.1
723	Hammonia	11.1
L724	Hapag	
725	Amanda	12.4
726	Joëlla	12.0
727	Nipponia	11.1
728	Leonisis	13.9
729	Watsonia	10.5
730	Athanasia	14.8
731	Sorga	10.5
732	Tjilaki	11.9
733	Mocia	10.2
734	Benda	11.1
735	Marghanna	10.7
736	Harvard	12.5
737	Arequipa	9.9
738	Alagasta	11.1
739	Mandeville	9.8
740	Cantabia	10.2
741	Botolphia	11.6
742	Edisona	10.7
743	Eugenisis	11.3
744	Aguntina	11.2
745	Mauritia	11.0
746	Marlu	10.8
747	Winchester	8.8
748	Simeïsa	9.9
749	Malzovia	12.9
750	Oskar	13.1
751	Faïna	9.8
752	Sulamitis	11.4
753	Tiflis	11.7
754	Malabar	10.3
755	Quintilla	10.8
756	Lilliana	11.3
757	Portlandia	11.4
758	Mancunia	9.6
759	Vinifera	11.7
760	Massinga	9.7
761	Brendelia	11.9
762	Pulcova	9.7
763	Cupido	13.8
764	Gedania	10.6
765	Mattiaca	14.1
766	Moguntia	11.0
767	Bondia	11.1
768	Struveana	11.4
769	Tatjana	10.0
770	Bali	12.0
771	Libera	11.8
772	Tanete	9.8
773	Irmintraud	10.6
774	Armor	10.0
775	Lumière	11.4
776	Berbericia	8.7
777	Gutemberga	11.1
778	Theobalda	10.6
779	Nina	9.8
780	Armenia	10.2
781	Kartvelia	10.6
782	Montefiore	12.7
783	Nora	12.1
784	Pickeringia	10.3
785	Zwetana	10.7
786	Bredichina	9.9
787	Moskva	11.4
788	Hohensteina	9.3
789	Lena	12.3
790	Pretoria	9.1
791	Ani	10.5
792	Metcalfia	11.2
793	Arizona	11.3
794	Irenaea	12.4
795	Fini	11.0
796	Sarita	10.2
797	Montana	11.8
798	Ruth	10.7
799	Gudula	11.5
800	Kressmannia	12.6
801	Helwerthia	12.2
802	Epyaxa	13.7
803	Picka	10.8
804	Hispania	8.9
805	Hormuthia	10.7
806	Gyldenia	11.2
807	Ceraskia	11.9
808	Merxia	10.8
809	Lundia	13.2
810	Atossa	14.2
811	Nauheima	11.9
812	Adele	12.5
813	Baumeia	13.3
814	Tauris	10.0
815	Coppelia	12.0
816	Juliana	11.4
817	Annika	12.0
818	Kapteynia	10.5
819	Barnardiana	13.3
820	Adriana	11.5
821	Fanny	12.6
822	Lalage	12.7
823	Sisigambis	12.6
824	Anastasia	11.3
825	Tanina	13.0
826	Henrika	12.8
827	Wolfiana	13.8
828	Lindemannia	11.3
829	Academia	12.0
830	Petropolitana	10.5
831	Stateira	13.6
832	Karin	12.3
833	Monica	12.3
834	Burnhamia	10.5
835	Olivia	12.3
836	Jole	14.4
837	Schwarzschilda	13.1
838	Seraphina	11.3
839	Valborg	11.9
840	Zenobia	10.6
841	Arabella	14.1
842	Kerstin	11.8
843	Nicolaia	14.8
844	Leontina	10.8
845	Naëma	11.6
846	Lipperta	11.4
847	Agnia	11.5
848	Inna	12.4
849	Ara	9.3
850	Altona	10.7
851	Zeissia	13.0
852	Wladilena	11.2
853	Nansenia	12.6
854	Frostia	13.6
855	Newcombia	13.1
856	Backlunda	11.9
857	Glasenappia	12.4
858	El Djezaïr	11.3
859	Bouzaréah	11.1
860	Ursina	10.8
861	Aïda	11.0
862	Franzia	11.3
863	Benkoela	10.3
864	Aase	14.3
865	Zubaida	13.4
866	Fatme	10.7
867	Kovacia	12.4
868	Lova	11.3
869	Mellena	13.4
870	Manto	13.0
871	Amneris	13.8
872	Holda	11.1
873	Mechthild	12.3
874	Rotraut	11.0
875	Nymphe	12.9
876	Scott	12.1
877	Walküre	11.9
L878	Mildred	
879	Ricarda	12.8
880	Herba	13.1
881	Athene	13.6
882	Swetlana	11.7
883	Matterania	13.9
884	Priamus	10.0
885	Ulrike	11.9
886	Washingtonia	9.7
887	Alinda	15.1
888	Parysatis	10.8
889	Erynia	12.7
890	Waltraut	11.9
891	Gunhild	11.3
892	Seeligeria	10.8
893	Leopoldina	10.9
894	Erda	11.0
895	Helio	9.5
896	Sphinx	13.0
897	Lysistrata	12.2
898	Hildegard	13.4
899	Jokaste	11.2
900	Rosalinde	12.9
901	Brunsia	13.2
902	Probitas	13.6
903	Nealley	10.8

Data for the Asteroids listed below includes Number, Name and Absolute Magnitude:

904	Rockefellia	11.4
905	Universitas	13.0
906	Repsolda	10.8
907	Rhoda	10.7
908	Buda	12.1
909	Ulla	9.7
910	Anneliese	11.4
911	Agamemnon	9.0
912	Maritima	10.4
913	Otila	13.8
914	Palisana	10.1
915	Cosette	13.1
916	America	12.7
917	Lyka	12.6
918	Itha	12.1
919	Ilsebill	12.5
920	Rogeria	12.4
921	Jovita	11.2
922	Schlutia	13.1
923	Herluga	12.7
924	Toni	10.5
925	Alphonsina	9.6
926	Imhilde	11.4
927	Ratisbona	10.4
928	Hildrun	10.8
929	Algunde	13.6
930	Westphalia	12.6
931	Whittemora	10.3
932	Hooveria	11.1
933	Susi	13.3
934	Thüringia	10.8
935	Clivia	14.3
936	Kunigunde	11.1
937	Bethgea	12.9
938	Chlosinde	12.4
939	Isberga	13.2
940	Kordula	10.3
941	Murray	12.6
942	Romilda	11.6
943	Begonia	10.8
944	Hidalgo	11.9
945	Barcelona	11.2
946	Poësia	11.5
947	Monterosa	11.4
948	Jucunda	12.6
949	Hel	10.9
950	Ahrensa	12.5
951	Gaspra	13.0
952	Caia	10.3
953	Painleva	11.6
954	Li	11.0
955	Alstede	12.7
956	Elisa	13.5
957	Camelia	11.0
958	Asplinda	11.7
959	Arne	11.9
960	Birgit	14.3
961	Gunnie	12.6
962	Aslög	12.8
963	Iduberga	13.8
964	Subamara	12.1
965	Angelica	11.4
966	Muschi	1.2
967	Helionape	13.7

968	Petunia	11.5
969	Leocadia	13.6
970	Primula	13.4
971	Alsatia	11.3
972	Cohnia	10.7
973	Aralia	11.0
974	Lioba	11.6
975	Perseverantia	11.5
976	Benjamina	10.5
977	Philippa	10.8
978	Aidamina	10.7
979	Ilsewa	11.0
980	Anacostia	9.2
981	Martina	12.2
982	Franklina	11.5
983	Gunila	10.9
984	Gretia	10.7
985	Rosina	14.2
986	Amelia	10.6
987	Wallia	10.7
988	Appella	12.5
989	Schwass-mannia	13.4
990	Yerkes	12.9
991	McDonalda	12.0
992	Swasey	12.1
993	Moultona	13.8
994	Otthild	11.4
995	Sternberga	11.5
996	Hilaritas	11.7
997	Priska	12.7
998	Bodea	12.2
999	Zachia	12.1
1000	Piazzia	11.4
1001	Gaussia	10.7
1002	Olbersia	12.1
1003	Lilofee	11.5
1004	Belopolskya	10.8
1005	Arago	11.0
1006	Lagrangea	12.9
1007	Pawlowia	12.6
1008	La Paz	11.7
1009	Sirene	15.3
1010	Marlene	11.8
1011	Laodamia	14.2
1012	Sarema	13.1
1013	Tombecka	11.0
1014	Semphyra	12.9
1015	Christa	10.1
1016	Anitra	13.3
1017	Jacqueline	12.3
1018	Arnolda	12.3
1019	Strackea	14.2
1020	Arcadia	12.2
1021	Flammario	10.1
1022	Olympiada	11.3
1023	Thomana	10.9
1024	Hale	11.9
1025	Riema	14.1
L1026	Ingrid	
1027	Aesculapia	11.9
1028	Lydina	10.4
1029	La Plata	12.1
1030	Vitja	11.6

1031	Arctica	10.6
1032	Pafuri	11.0
1033	Simona	12.3
1034	Mozartia	13.6
1035	Amata	11.8
1036	Ganymed	10.6
1037	Davidweilla	14.0
1038	Tuckia	11.7
1039	Sonneberga	12.4
1040	Klumpkea	11.2
1041	Asta	11.2
1042	Amazone	11.4
1043	Beate	10.9
1044	Teutonia	12.2
1045	Michela	14.3
1046	Edwin	11.6
1047	Geisha	13.4
1048	Feodosia	10.7
1049	Gotho	11.8
1050	Meta	13.9
1051	Merope	11.1
1052	Belgica	13.3
1053	Vigdis	13.6
1054	Forsytia	11.6
1055	Tynka	12.8
1056	Azalea	12.8
1057	Wanda	12.3
1058	Grubba	13.0
1059	Mussorgskia	11.8
1060	Magnolia	14.4
1061	Paeonia	12.0
1062	Ljuba	11.3
1063	Aquilegia	12.6
1064	Aethusa	12.3
1065	Amundsenia	13.5
1066	Lobelia	14.2
1067	Lunaria	12.0
1068	Nofretete	12.4
1069	Planckia	10.5
1070	Tunica	12.1
1071	Brita	11.3
1072	Malva	12.0
1073	Gellivara	12.6
1074	Belijawskya	11.3
1075	Helina	11.4
1076	Viola	13.0
1077	Campanula	14.0
1078	Mentha	12.7
1079	Mimosa	12.1
1080	Orchis	13.6
1081	Reseda	12.2
1082	Pirola	11.7
1083	Salvia	14.0
1084	Tamariwa	11.8
1085	Amaryllis	10.9
1086	Nata	10.8
1087	Arabis	11.0
1088	Mitaka	12.7
1089	Tama	12.9
1090	Sumida	14.9
1091	Spiraea	11.9
1092	Lilium	11.5
1093	Freda	10.0
1094	Siberia	13.4

1095	Tulipa	11.4
1096	Reunerta	11.4
1097	Vicia	12.9
1098	Hakone	11.8
1099	Figneria	11.5
1100	Arnica	12.4
1101	Clematis	11.9
1102	Pepita	10.5
1103	Sequoia	13.6
1104	Syringa	13.9
1105	Fragaria	11.1
1106	Cydonia	13.0
1107	Lictoria	10.4
1108	Demeter	12.4
1109	Tata	11.0
1110	Jaroslawa	13.3
1111	Reinmuthia	11.6
1112	Polonia	11.4
1113	Katja	10.7
1114	Lorraine	11.0
1115	Sabauda	10.5
1116	Catriona	10.8
1117	Reginita	13.3
1118	Hanskya	11.0
1119	Euboea	12.4
1120	Cannonia	13.4
1121	Natascha	12.6
1122	Neith	13.2
1123	Shapleya	12.9
1124	Stroobantia	12.0
1125	China	14.3
1126	Otero	13.1
1127	Mimi	12.1
1128	Astrid	12.0
1129	Neujmina	11.0
1130	Skuld	13.3
1131	Porzia	15.4
1132	Hollandia	12.3
1133	Lugduna	13.3
1134	Kepler	14.8
1135	Colchis	11.6
1136	Mercedes	12.2
1137	Raïssa	12.2
1138	Attica	12.2
1139	Atami	14.4
1140	Crimea	11.6
1141	Bohemia	14.6
1142	Aetolia	11.6
1143	Odysseus	9.5
1144	Oda	11.3
1145	Robelmonte	12.2
1146	Biarmia	11.0
1147	Stavropolis	13.2
1148	Raraju	11.5
1149	Volga	11.5
1150	Achaia	14.5
1151	Ithaka	14.9
1152	Pawona	12.3
1153	Wallenbergia	13.4
1154	Astronomia	11.5
1155	Aënna	12.9
1156	Kira	14.1
1157	Arabia	11.3
1158	Luda	12.2

1159	Granada	12.8
1160	Illyria	12.3
1161	Thessalia	12.8
1162	Larissa	10.6
1163	Saga	11.8
1164	Kobolda	14.4
1165	Imprinetta	11.8
1166	Sakuntala	12.7
1167	Dubiago	11.0
1168	Brandia	13.0
1169	Alwine	14.3
1170	Siva	13.1
1171	Rusthawelia	10.8
1172	Aeneas	9.3
1173	Anchises	10.2
1174	Marmara	12.9
1175	Margo	11.6
1176	Lucidor	12.4
1177	Gonnessia	10.2
1178	Irmela	13.0
L1179	Mally	
1180	Rita	10.2
1181	Lilith	12.7
1182	Ilona	12.6
1183	Jutta	13.0
1184	Gaea	12.3
1185	Nikko	13.4
1186	Turnera	10.4
1187	Afra	12.7
1188	Gothlandia	13.2
1189	Terentia	11.2
1190	Pelagia	13.3
1191	Alfaterna	11.7
1192	Prisma	13.7
1193	Africa	13.5
1194	Aletta	11.8
1195	Orangia	14.6
1196	Sheba	11.6
1197	Rhodesia	11.3
1198	Atlantis	16.0
1199	Geldonia	11.5
1200	Imperatrix	11.8
1201	Strenua	12.6
1202	Marina	11.4
1203	Nanna	13.2
1204	Renzia	13.3
1205	Ebella	15.3
1206	Numerowia	12.4
1207	Ostenia	12.1
1208	Troilus	9.9
1209	Pumma	11.4
1210	Morosovia	11.3
1211	Bressole	12.1
1212	Francette	8.0
1213	Algeria	12.1
1214	Richilde	12.1
1215	Boyer	12.5
1216	Askania	13.4
1217	Maximiliana	14.6
1218	Aster	14.2
1219	Britta	13.3
1220	Crocus	12.2
1221	Amor	19.2
1222	Tina	13.3

1223	Neckar	11.6
1224	Fantasia	12.8
1225	Ariane	13.7
1226	Golia	13.5
1227	Geranium	11.6
1228	Scabiosa	12.8
1229	Tilia	12.5
1230	Riceia	14.7
1231	Auricula	12.8
1232	Cortusa	11.4
1233	Korbresia	12.4
1234	Elyna	11.8
1235	Schorria	14.0
1236	Thaïs	12.9
1237	Geneviève	12.1
1238	Predappia	13.1
1239	Queteleta	13.8
1240	Centenaria	11.0
1241	Dysona	10.5
1242	Zambesia	11.4
1243	Pamela	11.1
1244	Deira	12.9
1245	Calvinia	11.0
1246	Chaka	12.0
1247	Memoria	11.7
1248	Jugurtha	11.0
1249	Rutherfordia	13.0
1250	Galanthus	14.4
1251	Hedera	11.8
1252	Celestia	12.3
1253	Frisia	13.3
1254	Erfordia	11.7
1255	Schilowa	11.6
1256	Normannia	10.9
1257	Móra	13.1
1258	Sicilia	11.7
1259	Ogyalla	11.9
1260	Walhalla	12.9
1261	Legia	11.9
1262	Sniadeckia	11.8
1263	Varsavia	11.7
1264	Letaba	10.9
1265	Schweikarda	11.1
1266	Tone	10.4
1267	Geertruida	13.7
1268	Libya	10.1
1269	Rollandia	9.8
1270	Datura	13.9
1271	Isergina	11.8
1272	Gefion	13.6
1273	Helma	14.0
1274	Delportia	13.3
1275	Cimbria	11.9
1276	Ucclia	11.9
1277	Dolores	12.3
1278	Kenya	12.2
1279	Uganda	13.8
1280	Baillauda	11.1
1281	Jeanne	12.7
1282	Utopia	11.3
1283	Komsomolia	12.0
1284	Latvia	11.4
1285	Julietta	11.6
1286	Bana-chiewicza	12.0
1287	Lorcia	12.2
1288	Santa	12.1

1289	Kutaïssi	11.6
1290	Albertine	13.7
1291	Phryne	11.5
1292	Luce	12.6
1293	Sonja	15.4
1294	Antwerpia	11.8
1295	Deflotte	11.6
1296	Andrée	12.5
1297	Quadea	12.5
1298	Nocturna	12.1
1299	Mertona	13.1
1300	Marcelle	12.4
1301	Yvonne	11.9
1302	Werra	11.9
1303	Luthera	10.5
1304	Arosa	10.4
1305	Pongola	11.6
1306	Scythia	10.8
1307	Cimmeria	13.5
1308	Halleria	11.9
1309	Hyperborea	11.4
1310	Villigera	12.8
1311	Knopfia	13.9
1312	Vassar	12.2
1313	Berna	13.0
1314	Paula	14.0
1315	Bronislawa	11.2
1316	Kasan	15.5
1317	Silvretta	11.1
1318	Nerina	13.2
1319	Disa	11.8
1320	Impala	12.0
1321	Majuba	11.5
1322	Copernicus	14.2
1323	Tugela	11.4
1324	Knysna	13.7
1325	Inanda	13.3
1326	Losaka	12.1
1327	Namaqua	13.3
1328	Devota	11.4
1329	Eliane	12.2
1330	Spridonia	11.3
1331	Solvejg	11.5
1332	Marconia	11.4
1333	Cevenola	12.9
1334	Lundmarka	11.2
1335	Demoulina	15.0
1336	Zeelandia	12.2
1337	Gerarda	12.1
1338	Duponta	14.1
1339	Désagneauxa	12.0
1340	Yvette	12.7
1341	Edmée	11.7
1342	Brabantia	13.4
1343	Nicole	12.6
1344	Caubeta	14.1
1345	Potomac	10.8
1346	Gotha	12.5
1347	Patria	12.2
1348	Michel	12.4
1349	Bechuana	11.7
1350	Rosselia	11.7
1351	Uzbekistania	11.1
1352	Wawel	12.4
1353	Maartje	11.1
1354	Botha	12.2
1355	Magoeba	13.9

1356	Nyanza	11.4
1357	Khama	12.1
1358	Gaika	13.0
1359	Prieska	11.7
1360	Tarka	12.5
1361	Leuschneria	12.0
1362	Griqua	12.4
1363	Herberta	12.8
1364	Safara	12.0
1365	Henyey	13.4
1366	Piccolo	11.6
1367	Nongoma	14.3
1368	Numidia	12.3
1369	Ostanina	11.9
1370	Hella	15.0
1371	Resi	12.4
1372	Haremari	12.8
1373	Cincinnati	14.3
1374	Isora	14.8
1375	Alfreda	12.9
1376	Michelle	13.8
1377	Roberbauxa	14.3
1378	Leonce	13.4
1379	Lomonosowa	12.1
1380	Volodia	13.2
1381	Danubia	13.0
1382	Gerti	13.4
1383	Limburgia	12.9
1384	Kniertje	12.9
1385	Gelria	12.1
1386	Storeria	14.8
1387	Kama	14.4
1388	Aphrodite	12.2
1389	Onnie	12.7
1390	Abastumani	10.2
1391	Carelia	13.7
1392	Pierre	12.9
1393	Sofala	13.3
1394	Algoa	12.9
1395	Aribeda	12.8
1396	Outeniqua	13.0
1397	Umtata	12.9
1398	Donnera	11.5
1399	Teneriffa	15.3
1400	Tirela	13.0
1401	Lavonne	13.5
1402	Eri	14.6
1403	Idelsonia	13.7
1404	Ajax	10.3
1405	Sibelius	14.4
1406	Komppa	12.5
1407	Lindelöf	12.4
1408	Trusanda	11.8
1409	Isko	11.8
1410	Margret	12.5
1411	Brauna	12.0
1412	Lagrula	13.5
1413	Roucarie	12.5
1414	Jérôme	13.8
1415	Malautra	13.6
1416	Renauxa	11.6
1417	Walinskia	12.3
1418	Fayeta	13.2
1419	Danzig	12.6
1420	Radcliffe	13.0
1421	Esperanto	11.5
1422	Strömgrenia	13.9

1423	Jose	12.5
1424	Sundmania	10.7
1425	Tuorla	12.9
1426	Riviera	12.1
1427	Ruvuma	11.9
1428	Mombasa	11.5
1429	Pemba	13.3
1430	Somalia	13.3
1431	Luanda	12.6
1432	Ethiopia	13.4
1433	Geramtina	12.9
1434	Margot	11.5
1435	Garlena	14.8
1436	Salonta	11.9
1437	Diomedes	9.4
1438	Wendeline	12.8
1439	Vogtia	11.6
1440	Rostia	12.8
1441	Bolyai	14.2
1442	Corvina	12.6
1443	Ruppina	12.5
1444	Pannonia	12.2
1445	Konkolya	11.9
1446	Sillanpää	14.0
1447	Utra	12.6
1448	Lindbladia	14.4
1449	Virtanen	13.8
1450	Raimonda	12.6
1451	Granö	13.9
1452	Hunnia	13.1
1453	Fennia	14.1
1454	Kalevala	13.9
1455	Mitchella	14.5
1456	Saldanha	12.1
1457	Ankara	12.5
1458	Mineura	12.4
1459	Magnya	11.4
1460	Haltia	13.8
1461	Jean-Jacques	10.9
1462	Zamenhof	12.2
1463	Nordenmarkia	12.1
1464	Armisticia	12.3
1465	Autonoma	12.2
1466	Mündleria	14.1
1467	Mashona	9.6
1468	Zomba	14.6
1469	Linzia	11.0
1470	Carla	11.7
1471	Tornio	12.5
1472	Muonio	13.8
1473	Ounas	13.6
1474	Beira	13.8
1475	Yalta	14.2
1476	Cox	14.9
1477	Bonsdorffia	12.8
1478	Vihuri	13.5
1479	Inkeri	12.8
1480	Aunus	14.4
1481	Tübingia	11.8
1482	Sebastiana	12.1
1483	Hakoila	12.6
1484	Postrema	12.3
1485	Isa	12.7
1486	Marilyn	14.5
1487	Boda	12.0
1488	Aura	12.0
1489	Attila	13.1

1490	Limpopo	13.3
1491	Balduinus	12.7
1492	Oppolzer	14.4
1493	Sigrid	12.4
1494	Savo	14.0
1495	Helsinki	13.5
1496	Turku	13.6
1497	Tampere	13.0
1498	Lahti	13.1
1499	Pori	12.7
1500	Jyväskylä	14.4
1501	Baade	13.6
1502	Arenda	12.6
1503	Kuopio	11.8
1504	Lappeenranta	13.0
1505	Koranna	12.6
1506	Xosa	13.2
1507	Vaasa	14.5
1508	Kemi	13.1
1509	Esclangona	14.1
1510	Charlois	12.5
1511	Daléra	14.2
1512	Oulu	10.5
1513	Mátra	14.1
1514	Ricouxa	13.6
1515	Perrotin	14.0
1516	Henry	13.3
1517	Beograd	12.2
1518	Rovaniemi	13.6
1519	Kajaani	12.4
1520	Imatra	11.5
1521	Seinäjoki	13.3
1522	Kokkola	13.7
1523	Pieksämäki	13.4
1524	Joensuu	11.9
1525	Savonlinna	13.2
1526	Mikkeli	14.8
1527	Malmquista	13.7
1528	Conrada	13.6
1529	Oterma	11.2
1530	Rantaseppä	14.3
1531	Hartmut	13.1
1532	Inari	12.0
1533	Saimaa	11.9
1534	Näsi	13.0
1535	Päijänne	12.9
1536	Pielinen	14.6
1537	Transylvania	12.9
1538	Detre	15.6
1539	Borrelly	12.2
1540	Kevola	11.9
1541	Estonia	12.7
1542	Schalén	11.7
1543	Bourgeois	13.6
1544	Vinterhansenia	12.9
1545	Thernöe	12.8
1546	Izsák	11.7
1547	Nele	12.0
1548	Palomaa	12.9
1549	Mikko	13.9
1550	Tito	13.4
1551	Argelander	13.7
1552	Bessel	12.7
1553	Bauersfelda	12.8
1554	Yugoslavia	12.7
1555	Dejan	12.7
1556	Wingolfia	11.4
1557	Roehla	12.3
1558	Järnefelt	11.5
1559	Kustaanheimo	13.0
1560	Strattonia	12.8
1561	Fricke	12.1
1562	Gondolatsch	13.1
1563	Noël	13.9
1564	Srbija	12.2
1565	Lemaître	13.8
1566	Icarus	17.7
1567	Alikoski	10.6
1568	Aisleen	13.2
1569	Evita	13.3
1570	Brunonia	12.6
1571	Cesco	13.2
1572	Posnania	11.3
1573	Väisälä	14.0
1574	Meyer	11.6
1575	Winifred	14.0
1576	Fabiola	11.9
1577	Reiss	15.3
1578	Kirkwood	11.6
1579	Herrick	11.1
1580	Betulia	15.8
1581	Abanderada	11.4
1582	Martir	13.1
1583	Antilochus	9.8
1584	Fuji	12.3
1585	Union	11.8
1586	Thiele	13.6
1587	Kahrstedt	12.9
1588	Descamisada	12.2
1589	Fanatica	13.3
1590	Tsiolkovskaja	13.3
1591	Baize	13.1
1592	Mathieu	12.8
1593	Fagnes	14.6
1594	Danjon	13.5
1595	Tanga	13.1
1596	Itzigsohn	11.6
1597	Laugier	13.4
1598	Paloque	14.4
1599	Giomus	12.2
1600	Vyssotsky	14.3
1601	Patry	13.9
1602	Indiana	13.9
1603	Neva	12.1
1604	Tombaugh	11.8
1605	Milankovitch	11.3
1606	Jekhovsky	12.9
1607	Mavis	12.8
1608	Muñoz	13.8
1609	Brenda	11.9
1610	Mirnaya	14.8
1611	Beyer	11.9
1612	Hirose	12.3
1613	Smiley	12.9
1614	Goldschmidt	11.6
1615	Bardwell	12.4
1616	Filipoff	12.3
1617	Alschmitt	12.1
1618	Dawn	13.0
1619	Ueta	12.5
1620	Geographos	16.7
1621	Druzhba	12.9
1622	Charcornac	13.5
1623	Vivian	11.9
1624	Rabe	12.0
1625	The NORC	11.7
1626	Sadeya	13.7
1627	Ivar	14.2
1628	Strobel	11.8
1629	Pecker	14.1
1630	Milet	12.6
1631	Kopff	13.3
1632	Sieböhme	12.5
1633	Chimay	11.8
1634	Ndola	14.6
1635	Bohrmann	12.8
1636	Porter	13.6
1637	Swings	11.3
1638	Ruanda	13.0
1639	Bower	12.0
1640	Nemo	14.7
1641	Tana	12.6
1642	Hill	12.6
1643	Brown	13.8
1644	Rafita	12.2
1645	Waterfield	12.7
1646	Rosseland	13.9
1647	Menelaus	11.6
1648	Shajna	13.1
1649	Fabre	12.8
1650	Heckmann	12.7
1651	Behrens	13.5
1652	Hergé	13.8
1653	Yakhontovia	12.8
1654	Bojeva	12.1
1655	Comas Solá	12.9
1656	Suomi	14.3
1657	Roemera	11.8
1658	Innes	13.0
1659	Punkaharju	11.3
1660	Wood	14.2
1661	Granule	14.1
1662	Hoffmann	13.1
1663	Van Den Bos	14.9
1664	Felix	13.8
1665	Gaby	12.4
1666	Van Gent	13.5
1667	Pels	13.6
1668	Hanna	13.6
1669	Dagmar	12.1
1670	Minnaert	12.3
1671	Chaika	13.0
1672	Gezelle	13.1
1673	Van Houten	12.2
1674	Groeneveld	12.0
1675	Simonida	13.2
1676	Kariba	14.2
1677	Tycho Brahe	13.4
1678	Hveen	12.1
1679	Nevanlinna	11.6
1680	Per Brahe	12.5
1681	Steinmetz	12.8
1682	Karel	14.6
1683	Castafiore	12.9
1684	Iguassú	12.8
1685	Toro	14.7
1686	De Sitter	12.0
1687	Glarona	11.5
1688	Wilkens	13.4
1689	Floris-Jan	12.9
1690	Maryhofer	11.9
1691	Oort	11.8
1692	Subbotina	12.5
1693	Hertzsprung	12.1
1694	Kaiser	13.7
1695	Walbeck	13.1
1696	Nurmela	14.4
1697	Koskenniemi	13.3
1698	Christophe	12.5
1699	Honkasalo	14.4
1700	Zvezdara	13.7
1701	Okavango	11.6
1702	Kalahari	12.3
1703	Barry	14.5
1704	Wachmann	13.9
1705	Tapio	14.3
1706	Dieckvoss	14.0
1707	Chantal	13.9
1708	Pólit	13.0
1709	Ukraina	14.0
1710	Gothard	14.6
1711	Sandrine	12.1
1712	Angola	11.1
1713	Bancilhon	14.5
1714	Sy	12.8
1715	Salli	15.5
1716	Peter	13.1
1717	Arlon	13.6
1718	Namibia	15.0
1719	Jens	12.6
1720	Niels	14.4
1721	Wells	12.0
1722	Goffin	13.1
1723	Klemola	11.2
1724	Vladimir	12.1
1725	CrAO	12.3
1726	Hoffmeister	13.1
1727	Mette	14.3
1728	Goethe Link	12.8
1729	Beryl	13.6
1730	Marceline	12.8
1731	Smuts	11.1
1732	Heike	12.0
1733	Silke	14.2
1734	Zhongolovich	12.6
1735	ITA	10.8
1736	Floirac	13.4
1737	Severny	12.2
1738	Oosterhoff	13.8
1739	Meyermann	13.8
1740	Paavo Nurmi	14.5
1741	Giclas	12.7
1742	Schaifers	12.4
1743	Schmidt	13.5
1744	Harriet	14.9
1745	Ferguson	13.2
1746	Brouwer	11.0
1747	Wright	14.6
1748	Mauderli	11.8
1749	Telamon	11.3
1750	Eckert	14.8
1751	Herget	13.6
1752	Van Herk	14.7
1753	Mieke	12.3
1754	Cunningham	10.8
1755	Lorbach	12.1
1756	Giacobini	14.0
1757	Porvoo	14.1
1758	Naantali	12.1
1759	Kienle	14.2
1760	Sandra	12.7
1761	Edmondson	12.7
1762	Russell	12.9
1763	Williams	14.3
1764	Cogshall	12.6
1765	Wrubel	11.1
1766	Slipher	13.6
1767	Lampland	13.4
1768	Appenzella	14.3
1769	Carlostorres	14.1
1770	Schlesinger	14.2
1771	Makover	11.3
1772	Gagarin	13.3
1773	Rumpelstilz	13.4
1774	Kulikov	13.6
1775	Zimmerwald	13.4
1776	Kuiper	12.1
1777	Gehrels	12.9
1778	Alfvén	12.9
1779	Paraná	15.4
1780	Kippes	12.0
1781	Van Biesbroeck	14.1
1782	Schneller	12.8
1783	Albitskij	12.7
1784	Benguella	13.6
1785	Wurm	14.0
1786	Raahe	12.2
1787	Chiny	12.5
1788	Kiess	12.9
1789	Dobrovolsky	14.4
1790	Volkov	14.1
1791	Patsayev	13.2
1792	Reni	13.1
1793	Zoya	13.8
1794	Finsen	12.0
1795	Woltjer	13.0
1796	Riga	11.7
1797	Schaumasse	14.0
1798	Watts	13.8
1799	Koussevitzky	12.5
1800	Aguilar	14.1
1801	Titicaca	12.4
1802	Zhang Heng	13.1
1803	Zwicky	13.4
1804	Chebotarev	13.5
1805	Dirikis	12.6
1806	Derice	14.1
1807	Slovakia	14.1
1808	Bellerophon	13.3
1809	Prometheus	12.8
1810	Epimetheus	13.9
1811	Bruwer	12.2
1812	Gilgamesh	12.7
1813	Imhotep	13.6
1814	Bach	14.3
1815	Beethoven	12.5
1816	Liberia	14.8
1817	Katanga	13.4
1818	Brahms	15.3
1819	Laputa	12.1
1820	Lohmann	14.7
1821	Aconcagua	14.9
1822	Waterman	14.7
1823	Gliese	14.2
1824	Haworth	12.9
1825	Klare	13.0
1826	Miller	12.4
1827	Atkinson	13.6
1828	Kashirina	12.3
1829	Dawson	13.8
1830	Pogson	13.8
1831	Nicholson	13.3
1832	Mrkos	11.7
1833	Shmakova	12.7
1834	1969 *QP*	12.8
1835	Gaidariva	12.8
1836	Komarov	12.7
1837	Osita	15.0
1838	Ursa	12.0
1839	Ragazza	12.8
1840	Hus	12.9
1841	Masaryk	11.8
1842	Hynek	13.6
1843	Jarmila	12.7
1844	Susilva	12.6
1845	Helewalda	13.0
1846	Bengt	14.6
1847	Stobbe	12.1
1848	Delvaux	11.9
1849	Kresák	12.3
1850	Kohoutek	14.3
1851	Lacroute	13.2
1852	Carpenter	11.9
1853	McElroy	11.7
1854	Skvortsov	13.7
1855	Korolev	13.9
1856	Růžena	13.5
1857	Parchomenko	13.8
1858	Lobachevskij	12.9
1859	Kovalevskaya	11.3
1860	Barbarossa	12.7
1861	Komenský	13.0
1862	Apollo	17.1
1863	Antinous	16.6
1864	Daedalus	16.3
1865	Cerberus	17.6
1866	Sisyphus	14.6
1867	Deiphobus	9.6
1868	Thersites	10.7
1869	Philoctetes	12.3
1870	Glaukos	12.0
1871	Astyanax	12.4
1872	Helenos	11.6
1873	Agenor	11.8
1874	Kacivelia	12.2
1875	1969 *QQ*	13.6
1876	Napolitania	16.1
1877	Marsden	12.5
1878	Hughes	12.5
1879	Broederstroom	14.2
1880	McCrosky	12.8
1881	Shao	12.2
1882	Rauma	12.0
1883	Rimito	14.4
1884	Skip	14.4
1885	Herero	14.8
1886	Lowell	12.8
1887	Virton	11.9
1888	Zu Chong-Zhi	13.2
1889	Pakhmutova	12.1
1890	Konoshenkova	12.6
1891	Gondola	13.1
1892	Lucienne	13.3
1893	Jakoba	12.5
1894	Haffner	13.5
1895	Larink	13.1
1896	Beer	14.9
1897	Hind	14.5
1898	Cowell	13.4
1899	Crommelin	14.1
1900	Katyuska	13.6
1901	Moravia	12.6
1902	Shaposhnikov	10.7
1903	Adzhimushkaj	12.1
1904	Massevitch	12.9
1905	Ambartsumian	14.1
1906	Naef	13.9
1907	Rudneva	13.3
1908	Pobeda	12.6
1909	Alekhin	13.5
1910	Mikhailov	11.8
1911	Schubart	11.4
1912	Anubis	13.1
1913	Sekanina	12.4
1914	Hartbeetspoortdan	13.7
1915	Quetzálcoatl	19.4
1916	Boreas	16.3
1917	Cuyo	16.6
1918	Aiguillon	12.4
1919	Clemence	15.1
1920	Sarmiento	15.7
1921	Pala	15.7
1922	Zulu	13.0
1923	Osiris	14.6
1924	Horus	14.3
1925	Franklin-Adams	13.6
1926	Demiddelaer	13.3
1927	Suvanto	13.0
1928	Summa	14.0
1929	Kollaa	13.6
1930	Lucifer	12.4
1931	1969 *QB*	14.5
1932	Jansky	14.7
1933	Tinchen	14.5
1934	Jeffers	14.1
1935	Lucerna	14.5
1936	Lugano	12.6
1937	Locarno	13.4
1938	Lausanna	14.1
1939	Loretta	12.1
1940	Whipple	12.6
1941	Wild	12.6
1942	Jablunka	14.3
1943	Anteros	16.6
1944	Gunter	14.9
1945	Wesselink	13.6
1946	1931 *PH*	14.1
1947	Iso-Heikkilä	11.6
1948	Kampala	13.6
1949	Messina	14.7
1950	Wempe	14.1
1951	Lick	17.3
1952	Hesburgh	11.4
1953	Rupertwildt	13.0
1954	Kukarkin	13.3
1955	McMath	12.7
1956	Artek	13.1
1957	Angara	12.1
1958	Chandra	12.2
1959	Karbyshev	14.1
1960	Guisan	12.7
1961	Dufour	12.4
1962	Dunant	13.4
1963	Bezovec	12.1
1964	Luyten	14.5
1965	Van De Kamp	13.4
1966	Tristan	15.1
1967	Menzel	14.1
1968	Mehltretter	12.9
1969	Alain	12.7
1970	1954 *ER*	12.8
1971	Hagihara	13.5
1972	Yi Xing	14.6
1973	Colocolo	13.1
1974	Caupolican	13.2
1975	Pikelner	13.3
1976	Kaverin	13.9
1977	Shura	12.5
1978	Patrice	14.3
1979	Sakharov	14.7
1980	Tezcatlipoca	15.5
1981	Midas	18.1
1982	Cline	13.9
1983	Bok	14.1
1984	Fedynskij	12.4
1985	Hopmann	12.4
1986	1935 SV$_1$	13.2
1987	Kaplan	13.0
1988	Delores	14.8
1989	Tatry	13.4
1990	Pilcher	14.1
1991	Darwin	14.7
1992	Galvarino	13.3
1993	Guacolda	13.6
1994	Shane	13.6
1995	Hajek	13.8
1996	Adams	13.3
1997	Leverrier	14.5
1998	Titius	12.9
1999	Hirayama	12.1
2000	Herschel	12.7
2001	Einstein	14.1
2002	Euler	13.4
2003	Harding	12.9
2004	Lexell	14.0
2005	Hencke	13.6
2006	Polonskaya	14.2
2007	McCuskey	13.1
2008	Konstitutsiya	11.3
2009	Voloshina	12.2
2010	Chebyshev	12.3
2011	Veteraniya	14.1
2012	Guo Shou-Jing	14.4
2013	Tucapel	13.3
2014	Vasilevskis	13.0
2015	Kacheuvskaya	13.5
2016	Heinemann	12.6
2017	Wesson	14.8
2018	Schuster	15.7
2019	1935 SX$_1$	13.6
2020	Ukko	11.8
2021	Poincaré	14.8
2022	West	12.4
2023	Asaph	12.8
2024	McLaughlin	14.5
2025	1953 *LG*	11.9
2026	Cottrell	14.6
2027	Shen Guo	12.9
2028	Janequeo	15.3
2029	Binomi	14.4
2030	Belyaev	14.8
2031	BAM	14.5
2032	Ethel	12.8
2033	Basilea	14.9
2034	Bernoulli	14.1
2035	Stearns	13.9
2036	Sheragul	14.1
2037	Tripaxeptalis	14.9
2038	Bistro	13.6
2039	Payne-Gaposchkin	14.1
2040	Chalonge	12.9
2041	Lancelot	13.6
2042	Sitarski	14.0
2043	Ortutay	12.2
2044	Wirt	14.4
2045	Peking	13.5
2046	Leningrad	12.2
2047	Smetana	15.1
2048	Dwornik	14.1
2049	Grietje	16.3
2050	Francis	13.1
2051	Chang	13.1
2052	Tamriko	11.1
2053	Nuki	12.8
2054	Gawain	13.6
2055	Dvořák	14.7
2056	Nancy	13.6
2057	Rosemary	14.0
2058	Róka	12.1
2059	Baboquivari	16.1
2060	Chiron	6.0
2061	Anza	18.1
2062	Aten	18.4
2063	Bacchus	18.8
2064	Thomsen	15.1
2065	Spicer	13.4
2066	Palala	14.2
2067	Aksnes	11.9
2068	Dangreen	12.9
2069	Hubble	12.4
2070	Humason	14.8
2071	Nadezhda	14.6
2072	Kosmodemyanskaya	13.3
2073	Janáček	13.9
2074	Shoemaker	15.0
2075	Martinez	15.1
2076	Levin	15.6
2077	Kiangsu	14.6
2078	Nanking	14.1
2079	Jacchia	13.4
2080	Jihlava	14.8
2081	Sázava	13.6
2082	Galahad	13.8
2083	Smither	14.3
2084	Okayama	13.6
2085	Henan	12.2
2086	Newell	13.2

Data for the Asteroids listed below includes Number, Name and Absolute Magnitude:

2087	Kochera	14.5
2088	Sahlia	14.3
2089	Cetacea	12.6
2090	Mizuho	11.5
2091	Sampo	12.0
2092	Sumiana	12.7
2093	Genichesk	14.3
2094	Magnitka	13.4
2095	Parsifal	13.9
2096	Väinö	14.3
2097	1953 *PV*	12.8
2098	Zyskin	13.4
2099	Opik	16.5
2100	Ra-Shalom	17.0
2101	Adonis	19.5
2102	Tantalus	17.5
2103	1960 *FL*	12.5
2104	Toronto	11.1
2105	Gudy	13.6
2106	Hugo	13.0
2107	Ilmari	13.0
2108	Otto Schmidt	12.7
2109	Dhotel	12.9
2110	Moore-Sitterly	14.8
2111	Tselina	11.3
2112	Ulyanov	13.8
2113	Ehrdni	13.4
2114	Wallenquist	12.5
2115	Irakli	12.5
2116	Mtskheta	13.5
2117	Danmark	13.0
2118	Flagstaff	12.0
2119	Schwall	14.9
2120	Tyumenia	11.8
2121	Sevastopol	13.7
2122	Pyatiletka	13.3
2123	Vltava	13.0
2124	Nissen	12.7
2125	Karl-Ontjes	13.0
2126	Gerasimovich	13.6
2127	Tanya	12.0
2128	Wetherill	15.2
2129	Cosicosi	15.2
2130	Evdokiya	15.0
2131	Mayall	13.5
2132	Zhukov	12.4
2133	Franceswright	14.5
2134	Dennispalm	14.3
2135	Aristaeus	19.2
2136	Jugta	12.8
2137	Priscilla	12.5
2138	Swissair	12.8
2139	Makharadze	13.5
2140	Kemerovo	12.2
2141	Simferopol	12.5
2142	Landau	12.9
2143	Jimarnold	15.3
2144	Marietta	12.4
2145	Blaauw	11.7
2146	Stentor	11.5
2147	Kharadze	13.0
2148	Epeios	12.0
2149	Schwambraniya	13.5
2150	1977 *TA*	15.0
2151	Hadwiger	12.0
2152	Hannibal	12.5
2153	Akiyama	13.0
2154	Underhill	13.8
2155	Wodan	13.6
2156	Kate	14.0
2157	Ashbrook	12.7
2158	1933 *OS*	12.6
2159	Kukkamäki	13.0
2160	Spitzer	13.7
2161	Grissom	13.4
2162	Anhui	14.0
2163	Korczak	12.8
2164	Lyalya	13.1
2165	Young	12.0
2166	Handahl	14.5
2167	Erin	12.9
2168	Swope	14.5
2169	Taiwan	13.3
2170	Byelorussia	14.7
2171	Kiev	15.0
2172	Plavsk	12.7
2173	Maresjev	12.6
2174	Asmodeus	14.5
2175	Andrea Doria	15.0
2176	Donar	13.4
2177	Oliver	12.9
2178	Kazakhstania	15.0
2179	Platzeck	13.0
2180	Marjaleena	12.0
2181	Fogelin	13.4
2182	Semirot	12.5
2183	1959 *OB*	12.6
2184	Fujian	12.0
2185	Guangdong	13.0
2186	Keldysh	13.5
2187	La Silla	14.0
2188	Orlenok	13.0
2189	Zaragoza	14.0
2190	Coubertin	13.6
2191	Uppsala	12.4
2192	Pyatigoriya	12.5
2193	Jackson	11.5
2194	Arpola	13.5
2195	Tengström	13.5
2196	Ellicott	11.0
2197	Shanghai	11.5
2198	Ceplecha	15.7
2199	Kleť	14.5
2200	Pasadena	13.9
2201	Oljato	16.7
2202	Pele	18.5
2203	1935 SQ$_1$	12.5
2204	Lyyli	13.0
2205	Glinka	12.9
2206	Gabrova	12.8
2207	Antenor	10.0
2208	Pushkin	11.5
2209	Tianjin	12.5
2210	Lois	15.6
2211	1951 WO$_2$	14.0
2212	Hephaistos	15.2
2213	Meeus	14.5
2214	Carol	12.9
2215	Sichuan	12.8
2216	Kerch	12.5
2217	Eltigen	12.2
2218	Wotho	12.9
2219	Mannucci	12.0
2220	Hicks	13.2
2221	Chilton	14.3
2222	Lermontov	12.5
2223	Sarpedon	9.8
2224	Tucson	13.1
2225	Serkowski	13.2
2226	Cunitza	13.5
2227	Otto Struve	15.0
2228	Soyuz-Apollo	12.5
2229	Mezzarco	14.0
2230	Yunnan	13.2
2231	Durrell	13.5
2232	Altaj	13.5
2233	Kuznetsov	14.0
2234	Schmadel	13.4
2235	Vittore	11.5
2236	Austrasia	13.5
2237	Melnikov	12.5
2238	Steshenko	13.0
2239	Paracelsus	13.0
2240	Tsai	13.0
2241	1979 *WM*	9.5
2242	1936 *TG*	14.5
2243	Lönnrot	14.0
2244	Tesla	13.4
2245	Hekatostos	12.5
2246	Bowell	11.8
2247	6512 *P—L*	14.8
2248	Kanda	13.0
2249	Yamamoto	12.0
2250	Stalingrad	12.5
2251	Tikhov	12.8
2252	CERGA	13.0
2253	Espinette	14.5
2254	Requiem	14.0
2255	Qinghai	12.5
2256	4519 *P—L*	13.1
2257	Kaarina	14.3
2258	Viipuri	13.0
2259	Sofievka	14.0
2260	Neoptolemus	10.0
2261	1977 *HC*	14.0
2262	Mitidika	13.5
2263	Shaanxi	12.5
2264	Sabrina	12.0
2265	Verbaandert	14.0
2266	Tchaikovsky	12.0
2267	Agassiz	14.5
2268	Szmytowna	13.0
2269	Efremiana	12.0
2270	Yazhi	12.2
2271	Kiso	12.0
2272	1972 *FA*	16.0
2273	Yarilo	14.0
2274	Ehrsson	14.5
2275	1979 *MH*	15.5
2276	Warck	14.0
2277	Moreau	13.7
2278	1953 *GE*	14.0
2279	Barto	14.0
2280	Kunikov	14.8
2281	1971 *UQ*$_1$	14.7
2282	Andrés Bello	15.0
2283	Bunke	13.7
2284	San Juan	14.0
2285	Ron Helin	15.0
2286	Fesenkov	14.5
2287	Kalmykia	14.5
2288	Karolinum	12.0
2289	6567 *P—L*	14.6
2290	1932 *CD*$_1$	13.5
2291	Kevo	11.5
2292	Seili	13.0
2293	Guernica	12.0
2294	1977 *PL*$_1$	12.6
2295	1977 *QD*$_1$	13.0
2296	1975 *BA*$_1$	12.5
2297	Daghestan	12.5
2298	Cindijon	16.0
2299	Hanko	14.5
2300	Stebbins	14.0
2301	Whitford	12.5
2302	1972 *TL*$_2$	13.0
2303	Retsina	13.5
2304	Slavia	13.5
2305	King	12.5
2306	1939 *PM*	13.0
2307	1957 *HJ*	12.1
2308	Schilt	13.5
2309	Mr. Spock	12.5
2310	Olshaniya	12.5
2311	El Leoncito	11.5
2312	Duboshin	11.0
2313	1976 *TA*	14.2
2314	Field	14.0
2315	Czechoslovakia	12.0
2316	Jo-Ann	13.6
2317	Galya	14.5
2318	Lubarsky	15.0
2319	7631 *P—L*	13.2
2320	1979 *QJ*	11.9
2321	Luznice	12.9
2322	Kitt Peak	14.0
2323	Zverev	12.0
2324	Janice	12.2
2325	Chernykh	12.5
2326	Tololo	12.0
2327	Gershberg	15.0
2328	Robeson	14.0
2329	Orthos	16.3
2330	Ontake	12.0
2331	Parvulesco	14.0
2332	Kalm	11.5
2333	Porthan	13.0
2334	Cuffey	14.5
2335	James	14.3
2336	Xinjiang	12.5
2337	1976 *UH*$_1$	12.5
2338	Bokhan	13.0
2339	2509 *P—L*	14.5
2340	Hathor	21.5
2341	Aoluta	14.0
2342	Lebedev	12.5
2343	Siding Spring	15.0
2344	Xizang	13.0
2345	Fučik	11.9
2346	Lilio	13.5
2347	1936 *TK*	12.5
2348	Michkovitch	13.5
2349	Kurchenko	12.5
2350	Von Lüde	14.5
2351	O'Higgins	14.5
2352	Kurchatov	12.2
2353	1975 *UD*	13.0
2354	Lavrov	13.0
2355	Nei Monggol	12.5
2356	Hirons	11.9
2357	Phereclos	10.0
2358	Bahner	12.0
2359	Debehogne	13.5
2360	Volgo-Don	13.5
2361	Gogol	13.0
2362	Mark Twain	15.0
2363	Cebriones	10.0
2364	Seillier	12.0
2365	Interkosmos	13.5
2366	Aaryn	15.0
2367	Praha	14.5
2368	Beltrovata	16.8
2369	Chekhov	13.0
2370	Van Altena	11.5
2371	Dimitrov	13.0
2372	Proskurin	13.0
2373	1929 *PC*	14.5
2374	Vladvysotskij	12.0
2375	1975 *AA*	12.0
2376	Martynov	12.0
2377	Shcheglov	13.0
2378	1935 *CY*	12.0
2379	Heiskanen	12.0
2380	Heilongjiang	14.0
2381	Landi	12.5
2382	Nonie	12.0
2383	Bradley	14.5
2384	Schulhof	13.5
2385	Mustel	14.5
2386	Nikonov	13.0
2387	1975 *FX*	13.0
2388	Gase	14.0
2389	Dibag	14.5
2390	Nežárka	13.0
2391	1957 *AA*	13.5
2392	Jonathan Murray	15.5
2393	Suzuki	11.8
2394	Nadeev	12.6
2395	Aho	13.6
2396	Kochi	12.5
2397	Lappajärvi	11.5
2398	Jilin	14.5
2399	Terradas	14.5
2400	Derevskaya	13.0
2401	Aehlita	13.5
2402	Satpaev	13.5
2403	Šumava	13.5
2404	Antarctica	13.0
2405	Welch	12.5
2406	Orelskaya	15.0
2407	1973 *DH*	12.0
2408	Astapovich	14.0
2409	Chapman	14.0
2410	Morrison	14.0
2411	Zellner	14.0
2412	Wil	13.0
2413	6816 *P—L*	13.0
2414	Vibeke	12.1
2415	1978 *UJ*	13.1
2416	Sharonov	12.2
2417	McVittie	13.0
2418	1971 *UV*	13.5
2419	Moldavia	14.5
2420	Čiurlionis	14.0
2421	Nininger	12.0
2422	Perovskaya	15.0
2423	Ibarruri	15.0
2424	Tautenburg	14.0
2425	1975 *FW*	13.0
2426	Simonov	12.5
2427	Kobzar	14.0
2428	Kamenyar	12.5
2429	1977 *TZ*	14.0
2430	Bruce Helin	13.5
2431	Skovoroda	13.5
2432	Soomana	14.0
2433	Sootiyo	13.0
2434	Bateson	12.0
2435	Horemheb	16.0
2436	Hatshepsut	13.5
2437	Amnestia	14.7
2438	Oleshko	14.3
2439	Ulugbek	12.5
2440	Educatio	15.0
2441	Hibbs	14.5
2442	Corbett	14.0
2443	Tomeileen	11.5
2444	Lederle	13.5
2445	Blazhko	14.0
2446	Lunacharsky	13.5
2447	Kronstadt	14.5
2448	1975 *BU*	13.0
2449	1978 *GC*	14.5
2450	Ioannisiani	12.5
2451	Dôllfus	13.0
2452	Lyot	13.0
2453	A921 *SA*	12.5
2454	Olaus Magnus	14.5
2455	Somville	13.0
2456	Palamedes	10.5
2457	1975 *TU*$_2$	14.0
2458	1977 *RC*$_7$	13.0
2459	Spellmann	13.0
2460	Mitlincoln	13.5
2461	1981 *EC*$_1$	12.5
2462	Nehalennia	15.0
2463	1934 *FF*	13.5
2464	Nordenskiöld	12.5
2465	1949 *PK*	13.5
2466	Golson	13.5
2467	Kollontai	14.0
2468	1969 *TO*$_1$	14.0
2469	Tadjikistan	13.0
2470	Agematsu	12.5
2471	Ultrajectum	13.0
2472	1973 *DG*	15.0
2473	Heyerdahl	14.0
2474	Ruby	13.0
2475	Semenov	12.3
2476	Andersen	12.2
2477	Biryukov	13.5
2478	Tokai	13.8
2479	Sodankylä	14.0
2480	1976 *YS*$_1$	15.0
2481	1977 *UQ*	14.5
2482	Perkin	12.5
2483	Guinevere	11.5
2484	Parenago	14.5
2485	1932 *BH*	13.0
2486	Metsähovi	13.5
2487	Juhani	14.0
2488	Bryan	15.0
2489	Suvorov	13.0
2490	Bussolini	13.0
2491	1977 *CB*	13.5
2492	Kutuzov	13.0
2493	Elmer	14.0
2494	Inge	12.0
2495	Noviomagum	16.5
2496	Fernandus	14.5
2497	Kulikovskij	14.1
2498	Tsesevich	13.1
2499	Brunk	13.1
2500	1926 *GC*	14.0
2501	Lohja	13.5
2502	Nummela	12.5
2503	Liaoning	15.5
2504	Gaviola	13.5
2505	Hebei	12.0
2506	Pirogov	13.0
2507	Bobone	13.0
2508	Alupka	15.0
2509	Chukotka	15.0
2510	Shandong	13.5
2511	Patterson	14.0
2512	Tavastiá	14.0
2513	Baetslé	14.0
2514	Taiyuan	13.5
2515	Gansu	14.0
2516	Roman	15.0
2517	1968 *SB*	13.0
2518	Rutllant	14.5
2519	Annagerman	12.0
2520	Novorossijsk	13.5
2521	1979 *DK*	13.0
2522	1980 *PP*	13.0
2523	1980 *PV*	13.0
2524	Budovicium	12.5
2525	O'Steen	12.0
2526	Alisary	13.5
2527	Gregory	14.5
2528	Mohler	12.8
2529	Rockwell Kent	14.3
2530	Shipka	13.0
2531	Cambridge	12.0
2532	Sutton	13.9
2533	A905 *VA*	12.5
2534	Houzeau	12.5
2535	Hämeenlinna	13.5
2536	Kozyrev	14.5
2537	Gilmore	13.5
2538	Vanderlinden	14.5
2539	Ningxia	15.5
2540	1971 *TH*$_2$	14.5
2541	1973 *DE*	13.5
2542	Calpurnia	12.5
2543	Machado	12.5
2544	Gubarev	14.0
2545	Verbiest	14.0
2546	1950 *FC*	13.5
2547	Hubei	14.5
2548	Leloir	13.5
2549	Baker	14.0
2550	Houssay	12.5
2551	Decabrina	13.0
2552	Remek	15.5
2553	Viljev	12.5
2554	Skiff	14.5
2555	Thomas	13.0
2556	Louise	15.0
2557	Putnam	13.5
2558	Viv	15.0
2559	1981 *UH*	13.5
2560	1932 *CW*	13.0
2561	Margolin	15.0
2562	1973 *FF*$_1$	12.5
2563	Boyarchuk	12.5
2564	1977 *QX*	14.5
2565	1977 *TB*$_1$	16.0
2566	Kirghizia	14.0
2567	Elba	12.5
2568	Maksutov	15.0
2569	Madeline	12.5
2570	Porphyro	13.0
2571	Geisei	15.5
2572	1950 *DL*	14.0
2573	Hannu Olavi	12.5
2574	Ladoga	13.0
2575	Bulgaria	14.5
2576	Yesenin	12.0
2577	Litva	14.0
2578	1975 *VW*$_3$	12.5
2579	Spartacus	14.5
2580	1977 *QP*$_4$	14.5
2581	1980 *VX*	15.0
2582	Harimaya-Bashi	12.0
2583	1975 *XA*$_3$	14.1
2584	Turkmenia	14.2
2585	Irpedina	13.8
2586	Matson	14.0
2587	Gardner	12.4
2588	Flavia	14.3
2589	Daniel	12.7
2590	Mourão	13.8
2591	1949 *PS*	12.5
2592	Hunan	12.0
2593	Buryatia	15.0
2594	1978 *TB*	12.5
2595	Gudiachvili	13.5
2596	Vainu Bappu	13.5
2597	Arthur	13.0
2598	Merlin	13.5
2599	Veselí	12.5
2600	Lumme	12.5
2601	Bologna	12.0
2602	Moore	14.0
2603	Taylor	13.0
2604	1972 *LD*$_1$	14.0
2605	Sahade	14.0
2606	Odessa	12.5
2607	Yakutia	15.0
2608	Seneca	18.8
2609	Kiril-Metodi	14.0
2610	Tuva	15.0
2611	Boyce	12.5
2612	Kathryn	12.0
2613	Plzeň	12.0
2614	Torrence	14.5
2615	1951 *RJ*	13.5
2616	Lesya	13.6
2617	Jiangxi	11.5
2618	Coonabarabran	13.4
2619	Skalnaté Pleso	13.8
2620	1980 *TN*	13.5
2621	Goto	11.8
2622	Bolzano	12.8
2623	A919 *SA*	14.5
2624	Samitchell	12.0
2625	Jack London	14.0
2626	Belnika	13.0
2627	Churyumov	13.0
2628	Kopal	14.0
2629	1980 *RB*$_1$	16.0
2630	1980 *TF*$_3$	13.0
2631	Zhejiang	13.0
2632	Guizhou	12.5
2633	Bishop	14.0
2634	James Bradley	11.5
2635	Huggins	14.0
2636	Lassell	12.5
2637	Bobrovnikoff	14.5
2638	Gadolin	13.5
2639	Planman	14.0
2640	Hällström	13.5
2641	1949 *GJ*	13.5
2642	1961 *RA*	14.0
2643	1973 *SD*	16.5
2644	Victor Jara	15.0
2645	Daphne Plane	13.5
2646	Abetti	12.5
2647	1980 *SP*	14.0
2648	Owa	14.0
2649	Oongaq	13.0
2650	1931 *EG*	12.5
2651	Karen	13.5
2652	1953 *GM*	13.0
2653	1964 *VP*	13.5
2654	Ristenpart	13.5
2655	Guangxi	12.5
2656	Evenkia	15.0
2657	Bashkiria	13.0
2658	Gingerich	13.0
2659	Millis	12.5
2660	Wasserman	13.5
2661	1982 *FC*$_1$	13.0
2662	Kandinsky	15.0
2663	6561 *P—L*	14.5
2664	Everhart	15.0
2665	1938 *DW*$_1$	14.5
2666	1951 *TA*	13.0
2667	1967 *UO*	13.5
2668	Tataria	14.5

2669	Shostakovich	13.5
2670	Chuvashia	12.0
2671	Abkhazia	15.0
2672	Pisek	14.5
2673	1980 *KN*	13.5
2674	Pandarus	10.0
2675	Tolkien	13.8
2676	Aarhus	13.9
2677	1935 *FF*	12.5
2678	Aavasaksa	13.5
2679	Kittisvaara	13.0
2680	1975 *NF*	14.5
2681	Ostrovskij	13.5
2682	Soromundi	15.0
2683	Brian	13.0
2684	Douglas	13.0
2685	Masursky	13.5
2686	Linda Susan	13.0
2687	1982 *HG*	13.0
2688	Halley	13.0
2689	1935 *CF*	14.5
2690	Ristiina	12.0
2691	1974 *KB*	14.5
2692	Chkalov	13.5
2693	Yan'an	14.5
2694	Pino Torinese	14.5
2695	Christabel	13.0
2696	Magion	13.0
2697	Albina	11.3
2698	Azerbajdzhan	13.0
2699	Kalinin	13.2
2700	Baikonur	13.3
2701	Cherson	13.5
2702	1978 SZ_2	12.7
2703	Rodari	14.8
2704	Julian Loewe	14.2
2705	Wu	14.9
2706	1980 *VW*	13.1
2707	Ueferji	12.8
2708	Burns	13.1
2709	Sagan	14.0
2710	Veverka	14.8
2711	Aleksandrov	12.7
2712	1937 *YD*	15.0
2713	1938 *EA*	12.5
2714	Matti	14.5
2715	Mielikki	13.0
2716	Tuulikki	14.5
2717	Tellervo	13.5
2718	1951 *OM*	13.0
2719	1965 *SU*	14.5
2720	Pyotr Pervyj	15.5
2721	Vsekhsvyatskij	13.0
2722	Abalakin	13.5
2723	Gorshkov	13.5
2724	Orlov	12.5
2725	David Bender	12.0
2726	Kotelnikov	13.5
2727	Paton	13.5
2728	Yatskiv	13.5
2729	1979 UA_2	13.0
2730	Barks	13.0
2731	Cucula	13.0
2732	Witt	13.0
2733	Hamina	14.0
2734	Hasek	12.5
2735	Ellen	14.5

2736	Ops	14.5
2737	Kotka	12.5
2738	1940 *EC*	13.2
2739	1952 UZ_1	14.4
2740	1974 SY_4	12.5
2741	Valdivia	12.8
2742	Gibson	13.2
2743	1965 *WR*	13.5
2744	Birgitta	16.5
2745	1976 SR_{10}	14.5
2746	Hissao	14.5
2747	1980 *DW*	12.5
2748	Patrick Gene	14.0
2749	1937 *TD*	12.5
2750	Loviisa	14.5
2751	1962 *RP*	14.0
2752	1965 *SP*	12.0
2753	1966 *DH*	13.0
2754	1966 *PD*	14.0
2755	Avicenna	13.5
2756	Dzhangar	14.5
2757	Crisser	12.0
2758	Cordelia	15.0
2759	Idomeneus	11.0
2760	Kacha	11.0
2761	Eddington	14.0
2762	Fowler	15.0
2763	Jeans	13.5
2764	Moeller	14.9
2765	1981 *EY*	12.9
2766	1982 FE_1	13.4
2767	1967 *UM*	12.5
2768	Gorky	14.0
2769	Mendeleev	12.5
2770	1977 SM_1	14.5
2771	1978 SP_7	13.5
2772	Dugan	14.5
2773	1981 JZ_2	14.5
2774	Tenojoki	12.0
2775	1953 TX_2	15.0
2776	Baikal	14.0
2777	Shukshin	14.0
2778	1979 *XP*	14.5
2779	Mary	14.5
2780	Monnig	14.0
2781	1982 *QH*	13.0
2782	2605 *P—L*	14.5
2783	Chernyshevskij	14.4
2784	Domeyko	14.3
2785	Sedov	13.0
2786	Grinevia	13.3
2787	Tovarishch	12.6
2788	1981 *EL*	14.2
2789	1956 *XA*	15.0
2790	1965 UU_1	12.5
2791	Paradise	13.0
2792	Ponomarev	14.5
2793	Valdaj	12.0
2794	Kulik	14.0
2795	Lepage	14.0
2796	Kron	13.5
2797	Teucer	9.5
2798	2009 *P—L*	14.0
2799	Justus	15.5
2800	4585 *P—L*	13.5
2801	1935 SU_1	13.0

2802	Weisell	12.0
2803	Vilho	13.0
2804	Yrjö	12.5
2805	Kalle	13.5
2806	1953 *GG*	13.5
2807	Karl Marx	13.5
2808	1976 *HS*	12.0
2809	Vernadskij	14.5
2810	Lev Tolstoj	14.0
2811	1980 *JA*	13.0
2812	Scaltriti	14.5
2813	Zappalà	12.5
2814	Vieira	13.0
2815	Soma	14.5
2816	Pien	13.0
2817	Perec	15.0
2818	2580 *P—L*	15.5
2819	1933 *UR*	13.5
2820	Iisalmi	14.0
2821	1978 *SQ*	15.0
2822	Sacajawea	13.5
2823	Van der Laan	14.5
2824	Franke	14.5
2825	1938 SD_1	14.0
2826	Ahti	12.0
2827	Vellamo	13.5
2828	Iku-Turso	14.0
2829	1948 *PK*	12.0
2830	Greenwich	13.5
2831	1930 *SZ*	13.9
2832	1975 EC_1	13.7
2833	Radishchev	13.2
2834	Christy Carol	13.2
2835	Ryoma	13.1
2836	Sobolev	12.3
2837	Griboedov	13.0
2838	1971 UM_1	15.0
2839	Annette	13.5
2840	Kallavesi	13.5
2841	Puijo	14.0
2842	1950 *OD*	13.0
2843	1975 *XQ*	14.0
2844	Hess	14.5
2845	Granklinken	14.5
2846	1942 *CJ*	11.8
2847	1959 CC_1	13.8
2848	1959 *VF*	12.5
2849	Shklovskij	13.7
2850	1978 TM_7	13.2
2851	1978 UQ_2	13.5
2852	1981 QU_2	13.4
2853	1963 *RG*	14.2
2854	1964 *JE*	14.2
2855	1931 TB_2	14.0
2856	1933 *GB*	12.5
2857	1942 *DA*	14.0
2858	1975 *XB*	15.5
2859	Paganini	14.0
2860	Pasacentennium	14.0
2861	Lambrecht	13.5
2862	Vavilov	14.0
2863	Ben Mayer	13.5
2864	Soderblom	14.0
2865	1935 *OK*	12.5
2866	1961 *TA*	13.0
2867	Steins	15.0

2868	1972 *UA*	14.5
2869	1980 RM_2	13.0
2870	Haupt	14.0
2871	Schober	14.0
2872	Gentelee	13.5
2873	Binzel	14.0
2874	Jim Young	15.0
2875	Lagerkvist	13.0
2876	Aeschylus	14.5
2877	Likhachev	13.5
2878	Panacea	12.5
2879	Shimizu	12.5
2880	Nihondaira	14.0
2881	1983 AA_1	15.0
2882	Tedesco	12.9
2883	Barabashov	14.5
2884	Reddish	13.0
2885	1939 *TC*	15.0
2886	1965 *YG*	14.5
2887	Krinov	14.0
2888	Hodgson	14.5
2889	1981 WT_1	12.7
2890	Vilyujsk	14.0
2891	McGetchin	12.5
2892	Filipenko	11.5
2893	1975 *QD*	10.0
2894	Kakhovka	13.5
2895	Memmon	10.5
2896	1931 *RN*	14.0
2897	Ole Rømer	14.0
2898	1938 *DN*	13.0
2899	1964 TR_2	14.0
2900	Lubos Perek	13.0
2901	1973 *DP*	13.0
2902	1980 FN_2	15.5
2903	1981 UV_9	13.0
2904	Millman	13.0
2905	Plaskett	13.0
2906	Caltech	11.0
2907	Nekrasov	12.5
2908	Shimoyama	13.0
2909	Hoshi-No-Ie	12.0
2910	Yoshkar-Ola	14.5
2911	1938 *GJ*	13.0
2912	1942 *DM*	13.5
2913	1931 *TK*	14.0
2914	1965 *SB*	15.5
2915	Moskvina	14.5
2916	Voronveliya	14.5
2917	Sawyer Hogg	13.0
2918	Salazar	13.5
2919	Dali	12.5
2920	Automedon	10.0
2921	Sophocles	14.5
2922	Dikan'ka	14.5
2923	Schuyler	14.0
2924	Mitake-Mura	13.0
2925	Beatty	15.0
2926	Caldeira	14.5
2927	Alamosa	13.5
2928	1976 GN_8	12.5
2929	Harris	12.5
2930	Euripides	14.0
2931	Mayakovsky	13.0
2932	Kempchinsky	13.0
2933	Amber	13.0
2934	Aristophanes	12.5

2935	1976 *UU*	14.0
2936	1979 *SF*	13.5
2937	Gibbs	14.0
2938	Hopi	12.5
2939	Coconino	14.0
2940	Bacon	15.0
2941	Alden	15.0
2942	1932 *BG*	14.0
2943	1933 *QU*	14.5
2944	1935 *QF*	14.0
2945	1935 ST_1	13.0
2946	1941 *UV*	14.0
2947	1955 QP_1	14.5
2948	Amosov	13.5
2949	Kaverznev	14.5
2950	1974 VQ_2	13.5
2951	1977 RB_8	11.5
2952	Lilliputia	15.5
2953	Vysheslavia	13.0
2954	Delsemme	15.0
2955	Newburn	14.5
2956	Yeomans	13.5
2957	1934 CB_1	11.5
2958	1981 *DG*	13.0
2959	Scholl	12.5
2960	Ohtaki	15.1
2961	Katsurahama	14.2
2962	1940 *YF*	12.5
2963	1964 VM_1	13.5
2964	1974 OA_1	13.5
2965	1975 *BX*	14.5
2966	1977 EB_2	15.0
2967	1977 SS_1	12.5
2968	1978 *QJ*	15.5
2969	1978 RU_1	14.0
2970	1978 *UC*	13.5
2971	1980 *YL*	14.5
2972	1939 *TB*	14.5
2973	1951 *AJ*	13.5
2974	1955 *QK*	15.0
2975	1970 AF_1	13.5
2976	Lautaro	12.0
2977	Chivilikhin	13.5
2978	1978 *SR*	13.0
2979	Murmansk	12.5
2980	1981 EU_{17}	14.0
2981	Chagall	13.0
2982	Muriel	13.0
2983	Poltava	13.0
2984	Chaucer	13.5
2985	Shakespeare	13.0
2986	Mrinalini	13.0
2987	Sarabhai	13.5
2988	1943 *EM*	13.0
2989	1976 UF_1	14.5
2990	1981 EN_{27}	14.5
2991	1982 *HV*	15.0
2992	Vondel	14.5
2993	1970 *PA*	13.5
2994	1975 *PA*	15.5
2995	Taratuta	14.0
2996	Bowman	13.0
2997	1974 *MJ*	15.0
2998	Berendeya	15.0
2999	Dante	14.5
3000	Leonardo	14.0
3001	Michelangelo	13.5

3002	1982 FB_3	14.0
3003	1983 *YH*	12.5
3004	1976 *DD*	16.0
3005	1979 QK_2	15.0
3006	Livadia	15.0
3007	Reaves	14.0
3008	Nijiri	13.0
3009	Coventry	15.5
3010	Ushakov	13.5
3011	1978 WM_{14}	13.0
3012	Minsk	12.0
3013	Dobrovoleva	14.5
3014	1979 *TM*	14.0
3015	Candy	12.5
3016	1981 *EK*	13.5
3017	1981 *UL*	13.5
3018	Godiva	14.0
3019	1940 *AC*	12.5
3020	1949 *PR*	13.0
3021	1967 *CB*	13.0
3022	1980 *SH*	15.0
3023	Heard	14.5
3024	1981 UW_9	12.0
3025	Higson	13.0
3026	1977 TA_1	13.0
3027	Shavarsh	14.5
3028	1978 TA_2	12.0
3029	1981 EA_8	14.5
3030	Vehrenberg	15.5
3031	Houston	14.0
3032	Evans	13.0
3033	Holbaek	14.0
3034	Climenhaga	13.5
3035	A924 *EJ*	13.5
3036	Krat	11.0
3037	1944 *BA*	13.0

3038	1978 QB_3	14.5
3039	Yangel	13.5
3040	Kozai	17.0
3041	Webb	13.5
3042	1981 EF_{10}	14.5
3043	San Diego	14.5
3044	1983 RE_3	13.0
3045	Alois	12.5
3046	Molière	13.5
3047	Goethe	14.5
3048	1964 TH_1	14.0
3049	1968 *FH*	12.5
3050	Carrera	15.0
3051	1974 *YP*	14.0
3052	Herzen	14.5
3053	Dresden	14.5
3054	Strugatskia	12.5
3055	1978 TR_3	13.5
3056	INAG	14.0
3057	Malaren	14.5
3058	Delmary	15.5
3059	1981 EF_{23}	15.0
3060	1982 RD_1	14.5
3061	Cook	13.0
3062	Wren	12.0
3063	Makhaon	10.0
3064	1984 BB_1	14.5
3065	1984 *CV*	13.5
3066	1984 *EO*	12.5
3067	1982 TE_2	14.5
3068	Khanina	14.5
3069	1982 UG_2	14.5
3070	1949 *GK*	15.0
3071	Nesterov	12.5
3072	Vilnius	15.0
3073	Kursk	14.5

3074	1979 YE_9	14.0
3075	1981 EY_{15}	15.0
3076	1982 RB_1	15.0
3077	Henderson	14.0
3078	Horrocks	12.5
3079	Schiller	14.5
3080	Moisseiev	13.0
3081	1971 *UP*	14.5
3082	Dzhalil	13.5
3083	1974 *MH*	15.5
3084	1977 QB_1	14.5
3085	1980 *DA*	14.0
3086	Kalbaugh	15.0
3087	Beatrice Tinsley	14.0
3088	1981 UX_9	13.0
3089	1981 XK_2	12.0
3090	Tjossem	13.5
3091	Van den Heuvel	15.0
3092	Herodotus	12.5
3093	1971 *MG*	12.5
3094	1979 FE_2	13.0
3095	1980 RT_2	12.5
3096	1981 QC_1	14.0
3097	Tacitus	13.0
3098	4579 *P—L*	16.0
3099	1940 *GF*	12.5
3100	1977 EQ_1	15.5
3101	Goldberger	15.0
3102	1981 *QA*	17.0
3103	1982 *BB*	16.0
3104	Dürer	12.5
3105	A907 *PB*	14.2
3106	Morabito	12.0
3107	Weaver	14.5

3108	1972 *QM*	15.0
3109	1974 *DC*	12.5
3110	1975 *SC*	14.5
3111	Misuzu	15.0
3112	1977 QC_5	14.5
3113	1978 *RO*	14.5
3114	Ercilla	14.5
3115	Baily	12.5
3116	Goodricke	13.5
3117	Niépce	13.0
3118	1974 *OD*	12.0
3119	Dobronravin	13.5
3120	1979 *RZ*	13.0
3121	1981 *EV*	14.5
3122	1981 ET_3	15.5
3123	Dunham	14.0
3124	Kansas	14.0
3125	Hay	13.5
3126	Davydov	12.5
3127	Bagration	13.5
3128	1979 FJ_2	12.5
3129	Bonestell	13.5
3130	1981 *YO*	14.0
3131	Mason-Dixon	13.0
3132	1940 *WL*	12.5
3133	Sendai	14.0
3134	Kostinsky	12.0
3135	1981 EC_9	15.0
3136	1981 WD_4	12.9
3137	1982 SM_1	14.6
3138	1980 *KL*	14.5
3139	1980 VL_1	12.0
3140	1983 *AO*	12.0
3141	1984 *RH*	11.5
3142	Kilopi	13.5
3143	1980 *UA*	14.0

Asteroids discovered after 1983 and issued names prior to 1987:

No	*Name*	*Issue*
3147	Samantha	Sept 86
3150	Tosa	June 86
3151	Talbot	March 86
3159	Prokofiev	June 86
3160	Anagerhofer	June 86
3165	Mikawa	Sept 85
3175	Netto	June 86
3181	Ahnert	July 85
3182	Shimanto	June 86
3190	Aposhanskij	Sept 86
3192	A'Hearn	June 86
3193	Elliot	June 86
3194	Dorsey	Dec 85
3197	Weissman	Sept 86
3198	Wallonia	June 86
3199	Nefertiti	Dec 85
3200	Phaethon	July 85
3215	Lapko	March 86
3216	Harrington	Sept 86
3217	Seidelmann	Sept 86
3219	Komaki	Sept 85
3220	Murayama	Dec 86
3224	Irkutsk	Sept 86
3225	Hoag	Dec 85
3227	Hasegawa	Dec 86
3237	Victorplatt	March 86
3245	Jensch	Sept 85
3249	Musashino	Dec 86
3254	Bus	July 85
3256	Daguerre	March 86
3262	Miune	June 86
3268	De Sanctis	June 86
3270	Dudley	Dec 85
3285	Ruth Wolfe	Dec 85
3288	Seleucus	Sept 85
3290	Azabu	Dec 86
3291	Dunlap	Sept 86
3292	Sather	Sept 86
3293	Rontaylor	Sept 86
3294	Carlvesely	Sept 86
3295	Murakami	Dec 86
3299	Hall	Dec 85
3310	Patsy	June 86
3312	Pederson	June 86
3317	Paris	Dec 85
3318	Blixen	Sept 86
3319	Kibi	Dec 86
3320	Namba	Dec 86
3333	Schaber	March 86
3338	Richter	March 86
3343	Nedzel	June 86
3344	Modena	March 86
3350	Scobee	March 86
3351	Smith	March 86
3352	McAuliffe	March 86
3353	Jarvis	March 86
3354	McNair	March 86
3355	Onizuka	March 86
3356	Resnik	March 86
3361	Orpheus	Dec 86
3362	Khufu	March 86
3367	Alex	March 86
3369	Freuchen	Sept 86
3379	Oishi	Dec 86
3383	Koyama	Dec 86
3391	Sinon	Dec 86
3392	Setouchi	Dec 86
3396	Muazzez	Dec 86
3415	Danby	Dec 86
3425	Huruka	Dec 86
3426	Seki	Dec 86
3431	Nakano	June 86
3432	Kobuchizawa	June 86
3449	Abell	Sept 86
3454	Lieske	Sept 86
3455	Kristensen	Sept 86

GEOGRAPHICAL INDEX

Ariel
(Moon of Uranus)

Feature	Coordinates
Abans	16°S, 251°E
Agape	47°S, 336°E
Ataksak	53°S, 225°E
Befanak	17°S, 32°E
Berylune	23°S, 328°E
Brownie Chasma	5-21°S, 325-357°E
Deive	23°S, 23°E
Djadek	12°S, 251°E
Domovoy	72°S, 339°E
Finvara	16°S, 19°E
Gwyn	78°S, 23°E
Huon	39°S, 33°E
Kachina Chasma	24-40°S, 210-280°E
Kewpie Chasma	15-42°S, 307-335°E
Korrigan Chasma	25-46°S, 328-353°E
Kra Chasma	32-36°S, 355-2°E
Laica	22°S, 44°E
Leprechaun Vallis	5-15°S, 350-25°E
Mab	39°S, 353°E
Melusine	53°S, 9°E
Onagh	22°S, 244°E
Pixie Chasma	18-25°S, 350-20°E
Rima	18°S. 260°E
Sprite Vallis	12-17°E, 332-355°E
Sylph Chasma	45-50°S, 328-15°E
Yangoor	68°S. 260°E

Callisto
(Moon of Jupiter)

Feature	Coordinates
Adal	77°N, 79°W
Adlinda	58°N, 20°W
Ägröi	42°N, 12°W
Akycha	74°N, 325°W
Alfr	9°S, 222°W
Ali	57°N, 58°W
Anarr	43°N, 3°W
Aningan	51°N, 11°W
Asgard	30°N, 140°W
Askr	53°N, 327°W
Balkr	27°N, 12°W
Bavorr	48°N, 23°W
Beli	61°N, 79°W
Bragi	77°N, 69°W
Brami	26°N, 18°W
Bran	25°S, 207°W
Buga	22°N, 326°W
Buri	43°S, 44°W
Burr	40°N, 136°W
Dag	56°N, 74°W
Danr	61°N, 75°W
Dia	73°N, 56°W
Dryops	77°N, 29°W
Durinn	66°N, 87°W
Egdir	31°N, 35°W
Erlik	66°N, 358°W
Fadir	56°N, 15°W
Fili	65°N, 349°W
Finnr	14°N, 14°W
Freki	82°N, 10°W
Frodi	69°N, 136°W
Fulla	74°N, 102°W
Fulnir	58°N, 37°W
Geri	66°N, 353°W
Gipul Catena	65°N, 55°W
Gisl	56°N, 35°W
Gloi	48°N, 246°W
Goll	58°N, 323°W
Gondul	59°N, 115°W
Grimr	43°N, 214°W
Gunnr	64°N, 100°W
Gymir	61°N, 55°W
Habrok	77°N, 129°W
Haki	26°N, 315°W
Har	6°N, 357°W
Hepti	64°N, 27°W
Hodr	69°N, 87°W
Hoenir	36°S, 261°W
Hogni	14°S, 5°W
Igaluk	5°N, 315°W
Ivarr	6°S, 322°W
Jumo	62°N, 15°W
Kari	47°N, 103°W
Karl	56°N, 335°W
Lodurr	52°S, 270°W
Loni	4°S, 215°W
Losy	68°N, 329°W
Mera	63°N, 73°W
Mimir	30°N, 54°W
Mitsina	57°N, 97°W
Modi	67°N, 115°W
Nama	57°N, 336°W
Nar	4°S, 45°W
Nerivik	22°S, 55°W
Nidi	66°N, 93°W
Nori	46°N, 347°W
Nuada	62°N, 269°W
Oski	56°N, 266°W
Ottar	60°N, 100°W
Pekko	17°N, 6°W
Reginn	42°N, 88°W
Rigr	69°N, 240°W
Sarakka	8°S, 53°W
Seqinek	55°N, 27°W
Sholmo	52°N, 18°W
Sigyn	33°N, 27°W
Skoll	57°N, 317°W
Skuld	6°N, 37°W
Sudri	53°N, 137°W
Sumbur	69°N, 332°W
Tindr	5°S, 355°W
Tornarsuk	25°N, 130°W
Tyn	68°N, 229°W
Valfodr	3°S, 246°W
Valhalla	10°N, 55°W
Vali	9°N, 327°W
Vestri	42°N, 54°W
Vitr	23°S, 347°W
Ymir	51°N, 97°W

Dione
(Moon of Saturn)

Feature	Coordinates
Aeneas	47°W, 26°N
Amata	287°W, 7°N
Anchises	63°W, 35°S
Antenor	8°W, 6°S
Caieta	80°W, 25°S
Carthage Linea	310-337°W, 10-20°N
Cassandra	245°W, 42°S
Catillus	275°W, 1°S
Coras	268°W, 3°N
Creusa	78°W, 48°N
Dido	15°W, 22°S
Ilia	344°W, 3°N
Italus	76°W, 20°S
Larissa Chasma	15-65°W, 20-48°N
Latagus	26°W, 16°N
Latium Chasma	64-75°W, 3-45°N
Lausus	23°W, 38°N
Magus	24°W, 20°N
Massicus	52°W, 36°S
Padua Linea	190-245°W, 5°N-40°S
Palatine Linea	285-320°W, 10-55°S
Remus	30°W, 10°S
Ripheus	29°W, 56°S
Romulus	24°W, 8°S
Sabinus	190°W, 44°S
Tibur Chasma	60-80°W, 48-80°N
Turnus	342°W, 21°N

Earth

Feature	Coordinates
Ahaggar Mountains	25°N, 6°E
Alps	42-48°N, 6-18°E
Amazon River	0-5°S, 50-72°E
Amur River	47-52°N, 126-141°E
Anadyr Mountains	70°N, 170°E
Andes Mountains	15°N-57°S, 65-81°W
Apennines Mountains	38-45°N, 10-15°E
Appalachian Mountains	33-48°N, 75-85°W
Aral Sea	45°N, 60°E
Arctic Ocean	80°N, 0-180°
Atacama Desert	18-57°S, 68-70°W
Atlantic Ocean	40°S-80°N, 75°W-15°E
Atlas Mountains	35°N, 10°E-10°W
Baltic Sea	10-23°E, 55-60°N
Barren Grounds	60-70°N, 95-120°W
Bering Sea	160-180°W, 50-65°N
Black Sea	41-47°N, 29-42°E
Blue Nile River	10-15°N, 31-39°E
Brahmaputra River	24-28°N, 90-95°E
Cape Horn	56°S, 67°W
Cape of Good Hope	34°S, 18°E
Caribbean Sea	2-21°N, 61-89°W
Carpathian Mountains	45-50°N, 22-26°E
Caspian Sea	36-47°N, 50-55°E
Caucasus Mountains	38-50°E, 40-45°N
Central Siberian Uplands	60-72°N, 85-120°E
Cerra Aconcagua	33°S, 69°
Challenger Deep	11°N, 141°E
Cherskiy Mountains	62-70°N, 135-150°E
Clarion Fracture Zone	10-20°N, 120-150°W
Clipperton Fracture Zone	5-10°N, 110-140°W
Coast Mountains	50-63°N, 123-138°W
Congo River	10°S-5°N, 12-30°E
Danube River	44-48°N, 9-29°E
Dead Sea	31-32°N, 35-36°E
Death Valley	36°N, 116°W
Don River	49-55°N, 38-45°E
East China Sea	22-35°N, 120-130°E
Euphrates River	31-37°N, 37-47°E
Ganges River	22-31°N, 78-91°E
Gobi Desert	40-50°N, 80-130°E
Grand Canyon	36°N, 112-113°W
Great Bear Lake	65-67°N, 117-125°W
Great Lakes	42-49°N, 76-92°W
Great Salt Lake	41-42°N, 112-113°W
Great Salt Lake Desert	41-42°N, 113-114°W
Great Slave Lake	61-63°N, 118-125°W
Gulf of Aden	10°S, 5-10°E
Gulf of Mexico	17-31°N, 83-97°W
Himalaya Mountains	27-38°N, 74-105°E
Kun Lun Shan	35-38°N, 80-95°E
Indian Ocean	7-55°S, 39-116°E
Indus River	24-29°N, 67-71°E
Irtysh River	52-67°N, 65-70°E
Kalahari Desert	23-30°S, 19-32°E
Kolyma Mountains	60-70°N, 150°E-180°
Lake Assal	13-15°N, 40-42°E
Lake Athabasca	58-60°N, 106-188°W
Lake Balkhash	46-48°N, 103-110°E
Lake Baykal	50-56°N, 103-110°E
Lake Chad	13-14°N, 14-15°E
Lake Erie	41-43°N, 78-84°W
Lake Eyre	27-30°S, 137°W
Lake Huron	43-46°N, 81-85°W
Lake Issyk-Kul	43°N, 76-78°E
Lake Koko Nor	36°N, 100°E
Lake Ladoga	60-62°N, 30-33°E
Lake Manitoba	50-52°N, 98-99°W
Lake Michigan	42-46°N, 85-88°W
Lake Nicaragua	11-12°N, 5°E-11°W
Lake Onega	61-63°N, 36°E
Lake Rudolf	3-5°N, 36°E
Lake Superior	46-49°N, 85-92°W
Lake Tanganyika	4-8°S, 29-31°E
Lake Titicaca	15°S, 69°W
Lake Torrens	30-33°S, 137°E
Lake Vanern	59°N, 12-19°E
Lake Victoria	1°N-3°S, 32-35°E
Lena River	56-73°N, 105-136°E
Libyan Desert	15-30°N, 20-32°E
Mackenzie River	61-69°N, 118-135°W
Madeira River	3-12°S, 58-71°W
Marcus-Necker Rise	20-25°N, 150°E-175°W
Mediterranean Sea	31-45°N, 6-31°E
Mekong River	9-21°N, 99-107°E
Mid-Atlantic Ridge	40°N, 30°W
Mississippi-Missouri River System	29-48°N, 89-113°W
Red Rock River System	89-113°W
Mojave Desert	34-36°N, 116-118°W
Mont Blanc	46°N, 7°E
Monte Rosa	46°N, 8°E
Mt Elbrus	44°N, 43°E
Mt Everest	28°N, 87°E
Mt Fujiyama	34°N, 138.5°E
Mt Illimani	10°N, 68°W
Mt K2 (Godwin Austen)	36°N, 76°E
Mt Kangchenjunga	27°N, 88°E
Mt Kenya	0°, 38°E
Mt Kilimanjaro	3°S, 37°E
Mt Logan	49°N, 67°W
Mt Meru	4°S, 37°E
Mt Ojos de Salado	27°S, 63.5°W
Mt Ras Dashan	13°N, 38°E
Mt Weisshorn	46°N, 8°E
Murray-Darling River System	30-39°S, 39-47°E
Namib Desert	15-30°S, 11-16°E
Niger River	5°E-11°W
Nile River	2-32°N, 30-34°E
Nubian Desert	20-22°N, 36-38°E
Nyasa Lake	21-24°S, 30°E
Ob River	51-67°N, 65-90°E
Ob-Irtysh River System	51-67°N, 65-90°E
Orinoco River	3-8°N, 61-68°W
Pacific Ocean	60°N-75°S, 120°W-70°E
Painted Desert	35-37°N, 110-112°W
Parana River	17-34°S, 61-47°W
Purus River	4-7°S, 61-67°W
Pyrenees	42-43°N, 2°W-3°E
Red Sea	11-31°N, 33-44°E
Rhine River	46-51°N, 3-10°E
Rio Grande River	26-32°N, 97-106°W
Rocky Mountains	20-70°N, 100-160°W
Sahara Desert	20-30°N, 15°W-20°E
St Lawrence River	44-70°N, 72-76°W
Salween River	17-32°N, 92-98°W
Sao Francisco River	8-16°S, 36-47°W
Sea of Japan	32-48°N, 128-142°E
Sea of Okhotsk	45-65°N, 135-165°E
South China Sea	0-25°N, 100-120°E
Spanish Sahara	23-28°N, 11-15°W
Takla Makan Desert	36-42°N, 75-95°E
Thar Desert	26-30°N, 70-75°E
Thoulette Deep	15°S, 69°W
Tien Shan Mountains	40-42°N, 70-88°E
Ural Mountains	51-68°N, 55-65°E
Vinson Massif	79°S, 86°W
Volga River	46-53°N, 45-54°W
Western Coastal Range	35-50°N, 119-124°W
Yangtze River	26-35°N, 91-112°E
Yellow River (Hwang Ho)	26-41°N, 87-119°E
Yellow Sea	30-41°N, 118-125°E
Yukon River	62-66°N, 136-163°W

Enceladus
(Moon of Saturn)

Feature	Coordinates
Ali Baba	11°W, 55°N
Bassorah Fossa	23-345°W, 40-50°N
Dalilah	244°W, 53°N
Daryabar Fossa	20-335°W, 5-10°N
Diyar Planitia	250°W, 0°
Dunyazad	200°W, 43°N
Harran Sulci	210-270°W, 5°S-35°N
Isbanir Fossa	0-350°W, 10°S-20°N
Julnar	340°W, 54°N
Samarkand Sulci	300-340°W, 10°S-75°N
Sarandib Planitia	300°W, 5°N
Shahrazad	200°W, 49°N
Shahryar	222°W, 58°N

Ganymede
(Moon of Jupiter)

Feature	Coordinates
Achlous	66°N, 4°W
Adad	62°N, 352°W
Adapa	83°N, 22°W
Ammura	36°N, 337°W
Anshar Sulcus	15°N, 200°W
Anu	68°N, 332°W
Apsu Sulci	40°S, 230°W
Aquarius Sulcus	50°N, 10°W
Asshur	56°N, 325°W
Aya	67°N, 303°W
Ba'al	29°N, 326°W
Barnard Regio	22°N, 10°W
Danel	4°N, 21°W
Dardanus Sulcus	20°S, 13°W
Diment	29°N, 346°W
Enlil	52°N, 301°W
Eshmun	22°S, 187°W
Etana	78°N, 310°W
Galileo Regio	35°N, 145°W
Gilgamesh	58°S, 124°W
Gula	68°N, 1°W
Harpagia Sulci	0°, 317°W
Hathor	70°S, 265°W
Isis	64°S, 197°W
Keret	22°N, 34°W
Khumbam	15°S, 332°W
Kishar	78°N, 330°W
Kishar Sulcus	15°S, 220°W
Marius Regio	10°S, 200°W
Mashu Sulcus	22°N, 200°W

Mimas

(Moon of Saturn)

Miranda

(Moon of Uranus)

Moon

(of Earth)

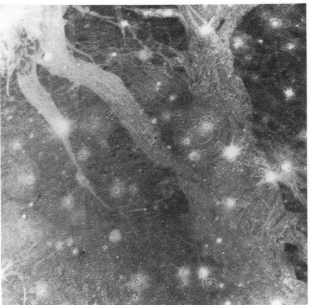

Above, from left: The Anshar, Mashu and Uruk Sulci etch their way across the icy, crater pocked face of Jupiter's moon Ganymede.

Oberon

(Moon of Uranus)

Rhea

(Moon of Saturn)

Tethys

(Moon of Saturn)

Titania

(Moon of Uranus)

GENERAL INDEX

GLOSSARY

Albedo: A measure of an object's reflecting power; the ratio of reflected light to incoming light in which complete reflection would give an albedo of 1.0.

Antoniadi Nomenclature: A universal system of designations devised by the Greek-born French astronomer Eugene (Eugenios) Antoniadi (1870–1944) for naming geologic features on solid surfaced celestial bodies other than the Earth or the Earth's Moon. Specifics of this system are included in this Glossary.

Aphelion: The point in an object's orbital path when it is farthest from the Sun. The opposite of Perihelion.

Astronomical Unit (AU): A unit of measurement used to calculate intra-Solar System distances. One AU is equal to the distance from the Earth to the Sun, or 93 million miles. The following are the mean distances from the Sun of the nine known planets:

planet	mean distance
Mercury	0.387 AU
Venus	0.723 AU
Earth	1.000 AU
Mars	1.524 AU
Jupiter	5.203 AU
Saturn	9.539 AU
Uranus	19.182 AU
Neptune	30.058 AU
Pluto	39.439 AU

Catena: Antoniadi nomenclature for a row of craters.

Chasma: Antoniadi nomenclature for a chasm or steep-sided canyon.

Conjunction: The alignment of two celestial bodies as viewed from a fixed point, such as from the Earth.

Dorsum: Antoniadi nomenclature for ridge.

Eccentricity: Eccentricity describes how elongated is an elliptic orbit; for ellipses $0 \leq$ eccentricity < 1. Eccentricity equals the distance between the foci divided by the major axis (which is twice the mean distance). An eccentricity of zero is a circle. For the planets, the mean distances and eccentricities are:

planet	mean distance	eccentricity
Mercury	0.387 AU	0.206
Venus	0.723 AU	0.007
Earth	1.000 AU	0.017
Mars	1.524 AU	0.093
Jupiter	5.203 AU	0.048
Saturn	9.539 AU	0.056
Uranus	19.182 AU	0.047
Neptune	30.058 AU	0.009
Pluto	39.439 AU	0.250

Ecliptic: The plane in which the Earth revolves around the Sun. All the planets except for Pluto (17 degrees) and Mercury (7 degrees) revolve around the Sun in planes that are within 3.4 degrees of the ecliptic, which is defined as zero degrees.

Ellipse: A geometrical shape such that the sum of the distances from any point on it to two fixed points (called the foci) is constant. All the planets have more or less elliptical orbits; a circle is a type of ellipse, but an elliptical orbit describes a planetary path that is more eccentric than concentric.

Elliptical orbit: An orbit that is not concentric, or circular, but that is shaped like an eccentric ellipse. All of the planets except Pluto have orbits that are very nearly perfectly circular. Pluto has an 'elliptical orbit.' Comets have very elliptical orbits.

Fossa: Antoniadi nomenclature for a narrow, shallow groove or ditch.

Labyrinthus: Antoniadi nomenclature for a labyrinth or a complex of interrelated canyons.

Light Year: A unit of measurement that equals the distance that light travels in one year at a speed of 186,281.7 miles per second.

Magnetosphere: The theoretically spherical region surrounding a star or a planet that is permeated by the magnetic field of that body.

Magnitude: The brightness of a star of other celestial body as viewed from Earth with the naked eye on a clear night. The scale ranges from Magnitude 1, the brightest, to Magnitude 6, the faintest.

Mare: A 'sea' as observed on Earth's Moon. It is actually a vast open basalt plateau and not a 'sea' in the sense of the Earth's seas. The plural is 'maria.'

Mensa: Antoniadi nomenclature for a mesa or butte.

Mons: Antoniadi nomenclature for a mountain, particularly but not limited to volcanic mountains.

Patera: Antoniadi nomenclature for an irregular, usually volcanic crater or caldera.

Perihelion: The point in an object's orbital path when it is closest to the Sun. The opposite of Aphelion.

Planitia: Antoniadi nomenclature for a lowland plateau or basin.

Planum: Antoniadi nomenclature for an upland plateau.

Protostellar: 'Pre-Star' (adjective). A term used in reference to the materials (hydrogen and helium) that will become a star, while they are still a 'pre-star' gas cloud.

Sidereal Period: For objects in the Solar System, the duration of time taken for a body to make a complete orbit or revolution around the Sun. This translates as that body's year. The Earth's sidereal period is 365.256 days. In a broader sense, a sidereal period is the orbital or rotation period of any object with respect to the fixed stars, or as seen by a distant observer.

Synodic Period: The orbital or rotational period of an object as seen by an observer on the Earth. For the Moon or a planet, the synodic period is the interval between repetitions of the same phase or configuration.

Tholus: Antoniadi nomenclature for an isolated hill or mountain.

Valles/Vallis: Antoniadi nomenclature for a valley.

Vastitas: Antoniadi nomenclature for a particularly vast planitia or planum.